# A Future Officer Career Management System

## *An Objectives-Based Design*

Harry J. Thie
Margaret C. Harrell
Roger A. Brown
Clifford M. Graf II
Mark Berends
Claire M. Levy
Jerry M. Sollinger

Prepared for the
Office of the Secretary of Defense

*National Defense Research Institute*

**RAND**

The research described in this report was sponsored by the Office of the Secretary of Defense (OSD). The research was conducted in RAND's National Defense Research Institute, a federally funded research and development center supported by the OSD, the Joint Staff, the unified commands, and the defense agencies under Contract DASW01-95-C-0059.

Library of Congress Cataloging-in-Publication Data

A future officer career management system : an objectives-based design / Harry J. Thie ... [et al.].
p. cm.
"Prepared for the Office of the Secretary of Defense."
"National Defense Research Institute."
"MR-788-OSD."
Includes bibliographical references.
ISBN 0-8330-2491-4 (alk. paper)
1. United States—Armed Forces—Officers.
2. United States—Armed Forces—Personnel management.
3. Career development. I. Thie, Harry J. II. United States. Dept. of Defense. Office of the Secretary of Defense.
III. National Defense Research Institute (U.S.). IV. RAND.
UB413.F88 2001
355.3' 31' 0973—dc21 97-6233
CIP



Published 2001 by RAND
1700 Main Street, P.O. Box 2138, Santa Monica, CA 90407-2138
1200 South Hayes Street, Arlington, VA 22202-5050
RAND URL: http://www.rand.org/
To order RAND documents or to obtain additional information, contact Distribution Services: Telephone: (310) 451-7002;
Fax: (310) 451-6915; Internet: order@rand.org

# PREFACE

This report represents a follow-on effort to an earlier National Defense Research Institute study, *Future Career Management Systems for U.S. Military Officers,* MR-470-OSD, 1994, that serves as a source for alternative career management practices. The current report applies an objectives-based methodology using preferences of Department of Defense policymakers in the development of a career management system for officers. The work was completed and provided to the sponsor in 1997.

This study was sponsored by the Under Secretary of Defense for Personnel and Readiness, and it was carried out in the Forces and Resources Policy Center of the National Defense Research Institute, a federally funded research and development center sponsored by the Office of the Secretary of Defense, the Joint Staff, the unified commands, and the defense agencies. The study should interest those involved with military personnel management.

# CONTENTS

# FIGURES

# TABLES

# SUMMARY

## BACKGROUND AND PURPOSE

The current officer management system was designed to meet the challenges of the cold war. With the disappearance of the United States' only global rival and the end of the cold war, members of Congress and senior leaders in the Defense Department began to question whether a career management system designed for the cold war could also serve the nation's security needs in the new environment. Some evidence suggests that it cannot. The current system, largely the result of the 1980 Defense Officer Personnel Management Act (DOPMA), did not prove itself a good management tool in the turbulent period following the collapse of the Berlin Wall. Furthermore, the changes in the size and missions of the armed forces suggest a need not so much for a different type of officer as for more different types.

In response to these concerns, RAND's National Defense Research Institute (NDRI) completed an earlier study on officer career management systems for the future (see MR-470-OSD). That study determined a likely range of future officer requirements, defined a number of alternative career management systems, and evaluated them. Its purpose was to provide policymakers a toolbox from which they could select needed policies to address goals for future careers. However, that study did not attempt to define a "best" system, because one of the key components for designing a system—the objectives it was to accomplish—was missing.

This study picks up where the previous one stopped. It designs a "best" officer career system, "best" being defined as the one that most fully satisfies the objectives of current policymakers. In so doing, it draws on the earlier study in three ways. First, it uses data from the earlier effort (e.g., the costs of acquiring an officer). Second, it accepts the four-category structure of the officer corps: line, specialist, support, and professional. **Line officers** have unique military skills, particularly those directly involved in combat operations and related military functions. **Specialists** practice any military skills that require recurring assignments and utilization of advanced education or high cost, long-duration training or experience. **Support officers** have skills analogous to civilian white-collar occupations needed to support military organizations where general military experience is desired or will assist task performance. **Professionals** have civilian professional skills not usually requiring significant military experience (e.g., medical, dental, legal, and chaplain). Third, it uses the description of an officer career system developed in MR-470-OSD: four interrelated personnel functions—accessing, developing, promoting, and transitioning.

**Accessing** pertains to how officers enter the system. Thus, the different sources of officers [e.g., academies, Reserve Officer Training Corps (ROTC), Officer Candidate School/Officers' Training School (OCS/OTS)]; when they enter into careers; and the educational, physical, aptitude, and moral standards that officers must meet at entry are all issues of accession.

**Developing** refers to the ways officers are moved through the career management system. The length and number of assignments in an officer's career; the nature of those assignments, often referred to as a career path; and the amount of education needed and obtained during a career are all issues determined by developing policies.

**Promoting** involves policies pertaining to promotion opportunity; the timing (time in service before promotion to particular grades or the time in each grade between promotions); and the basis for promotion (whether the emphasis is upon merit or seniority).

**Transitioning** has to do with the ways officers leave the system. These would include policies about when turnover should occur; tenure (limits on involuntary separation that protect the individual);

the minimum service required before an officer is vested for some future annuity payment; and the maximum service allowed before mandatory retirement.

## DESIGNING AN OFFICER CAREER MANAGEMENT SYSTEM

At the highest level, designing an officer career management system involves a relatively straightforward two-step process. The first step identifies the objectives of the system and establishes a relative priority among them. The second step determines the effect of those objectives on the four personnel areas.

### Identifying Objectives

To identify the objectives of an officer career management system, we drew on a wide range of sources. These included an extensive review of career management literature and multiple seminars with senior military and civilian leaders and members of the different military services. The process extended over three years and yielded a core of 11 objectives. Although all are clearly important, it is also clear that all are not equally important. So the next task was to establish a relative rank for the objectives. To accomplish that task, we resorted to a group of senior military and civilian officials who were deeply involved in career management issues. We used a rigorously structured interview technique that required the respondents to score their preferences between two objectives. With scores for various pairings of objectives, we were then able to rank the respondents' preferences. Averaging across all respondents, we then determined a ranking for all objectives. The objectives and their relative rankings as a percentage of 100 appear below.

### Determining the Effect of the Objectives

Determining how the objectives affect the four personnel areas is fairly complex. Each personnel function has a number of aspects (there are 17 in all), and each aspect has various alternatives (we designed 58 for this study). To illustrate, the point at which an officer begins a career—called entry point—is but one of four aspects of accessing. The others are type of pre-entry acculturation, amount of

| Objective | Percent Weight |
|---|---|
| Keep costs reasonable | 19 |
| Provide career satisfaction | 13 |
| Emphasize cadre with military culture | 13 |
| Meet active experience needs | 12 |
| Meet active skill needs | 11 |
| Inculcate culture prior to or at entry | 8 |
| Provide high opportunity to serve | 7 |
| Provide career opportunity | 6 |
| Meet reserve needs | 5 |
| Meet active grade needs | 3 |
| Be compatible with civilian careers | 2 |

obligated service, and length of initial tenure. There are at least three alternatives for entry point. Officers can enter at the beginning of a career as they do now, enter from prior active or reserve service, or enter from civilian life at some point along a career. Figure S.1 portrays the relationships among the three elements.

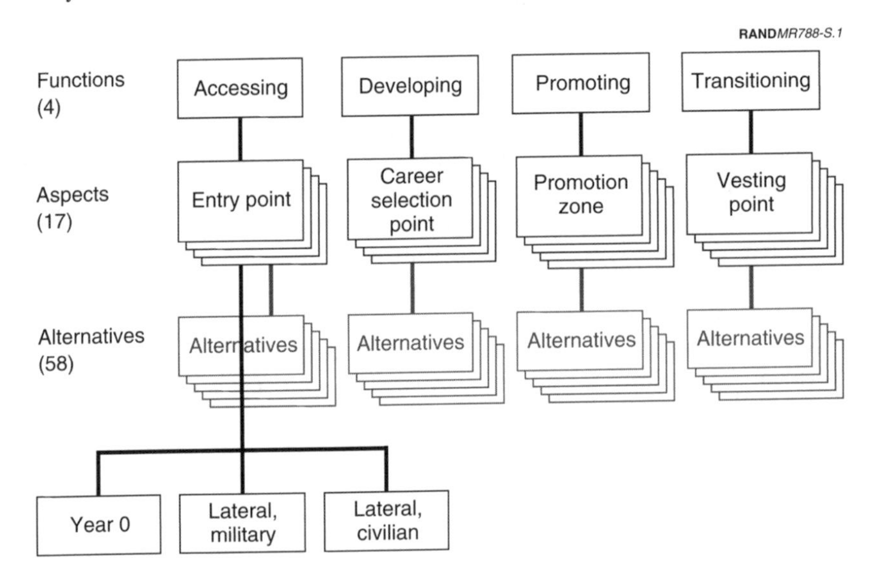


Figure S.1—Relationship Among Functions, Aspects, and Alternatives

It takes almost 750 decisions to determine the effect of the 11 objectives on the 17 aspects for four skill groups. The functions, aspects, and alternatives appear in Table S.1.

**Table S.1**

**Functions, Aspects, and Alternatives**

| Function | Aspects | Alternatives |
|---|---|---|
| Accessing | Entry point | Lateral from civilian<br>Lateral from military<br>Year 0 |
| | Initial tenure | 2 years<br>4 years<br>6 years |
| | Pre-entry acculturation | None<br>Educational, high-intensity, short<br>Educational, low-intensity, long<br>Educational, high-intensity, long<br>Experiential, medium intensity, medium length |
| | Amount of obligated service for education, training | 0.5 year<br>1 year<br>1.5 years<br>2 years |
| Developing | Career selection point | None<br>5–10 years<br>8–10 years<br>> 10 years |
| | Effect of nonselection | Separation<br>Migration to new skill |
| | Average assignment length | Decrease by two-thirds of average<br>Decrease by one-third of average<br>Current length<br>Increase by one-third of average<br>Increase by two-thirds of average |
| | Military and civilian education | Current amount<br>2 years more<br>2 years less |
| Promoting | Promotion zone | Time in service<br>Time in grade<br>Combination |

**Table S.1—continued**

| Function | Aspects | Alternatives |
|---|---|---|
| Promoting | Length of zone | Narrow (1–2 years)<br>Broad (3–8 years)<br>Open |
| | Opportunity | Fixed<br>Selective (based on requirements) |
| | Nature of continuation | Guaranteed<br>Based on requirements |
| Transitioning | Vesting point | 4–9 years<br>10–15 years<br>20 years |
| | Transitional ability of the system | Tenure<br>Voluntary separation incentives<br>Neither tenure nor incentives |
| | Maximum career length | 30 years<br>35 years<br>40 years |
| | Separation rates in first 10 years | High<br>Medium<br>Low |
| | Retirement annuity point | 15 years<br>20 years<br>25 years<br>30 years<br>35 years<br>40 years |

A rigorous methodology is needed to sort these many decisions. We use multiobjective decision analysis, which provides a powerful tool not only for evaluating large numbers of alternatives against numerous objectives but also for incorporating and retaining the preferences of the policymakers in the decision analysis process. Conceptually, the process determines how well alternatives achieve objectives. The alternatives are ranked and scored based on qualitative and quantitative data to determine how well they achieve each objective. The scores are normalized to enable comparison across objectives. The preferences (weights) for each objective are then applied to the alternatives' scores, which are summed to provide an overall ranking and score for each alternative. These separate decisions form the basis for larger decisions about which alternatives best meet all the objectives for each officer skill group.

Several points need to be made about the career management system that results from this methodology. First, the system reflects the judgment of a specific group of policymakers about what the objectives of that system should be and the relative preference of that group for different objectives. Second, the policymakers did not design the alternatives or know the effect of the ranking of their objectives on the alternatives until the conclusion of the study. Finally, the study considers the direct relation between objectives and alternatives but not the complex interactions that occur among them. Implementation of this system will require additional analysis. The officer career management system described here points to the likely future directions given the set of objectives.

## RESULTS

The career management system that results from our analysis reflects the objectives of a set of policymakers currently involved in military personnel management and our knowledge of the effects of the alternatives on the objectives. That is, these policymakers set objectives for the management system and established a relative priority among them. Should the priorities change significantly or new evidence emerge about the effect of alternatives, the system described here might no longer be appropriate.

The system that emerges from our analysis for line officers differs significantly from the current one in each of the personnel functions.

### Accessing

All methods of accession acculturate well, so a combination, as is now used, seems valid. However, enlisted service provides the best acculturation under this set of objectives, implying that greater use could be made of the enlisted force to fill the officer ranks.

Education and training should command a longer payback than they do now. We evaluated a maximum payback of 2 for 1, but the longer the payback period, the higher the alternative scored. Few accession programs now demand a 2-for-1 payback.

## Developing

Some mandatory skill transfers occur after six years. The line specialty typically requires more junior officers than the other specialties do, so after six years the line group needs to be trimmed to match the number of senior positions. Some officers who want to stay in the service will be transferred into the other specialty groups; others will be separated.

Officers would remain in assignments significantly longer. The increase could be up to two-thirds the length of a current assignment (e.g., moving from three-year to five-year assignments).

This system provides for more in-service education—an additional two years over the life of a full career.

## Promoting

Some of the most radical departures from the current system occur in the promotion personnel function. The current system tended to score lowest of all alternatives.

Promotions would occur based on time in grade, given a minimum time in service.

Promotion zones would be long, from three to eight years, compared with the current system's one-year zone. Furthermore, the system is requirements based, advancing officers only as vacancies exist. Officers would be selectively continued even though they were not selected for promotion, based on service needs.

## Transitioning

Considerable differences also exist between the alternatives in this function and those in the present system.

The system developed here has vesting in a future annuity occur somewhere between four and nine years of service. In the current system, no vesting is possible until 20 years, at which time an immediate annuity is available.

This system fosters high turnover early in careers but longer careers—up to 40 years. Also, it delays the availability of immediate annuities to within five years of a full career. Thus, while service members vest sooner than they do in the current system, they do not receive the benefit until considerably later.

These differences suggest that policymakers face a choice. If the objectives correctly reflect their desires for the system and their relative priorities, they should modify the current system. Otherwise, they should modify either their objectives or their assigned priorities and determine what changes the new set suggests. In either case, policymakers should be clear about objectives and preferences for them as a basis for officer career management.

# ACKNOWLEDGMENTS

The authors wish to thank the staff of the Deputy Assistant Secretary of Defense (Military Personnel Policy) for their support, particularly Colonel Thomas Page, Colonel Toreaser Steele, Colonel David Schreier, and Lieutenant Colonel Elaine Parker. Additionally, many officers and civilians representing all the military departments participated in the policymaking group and the working group established by the Assistant Secretary of Defense (Force Management Policy). These groups contributed their time, policymaking and analytical expertise, and their attitudes about the future. We would also like to thank Mr. Robert Emmerichs and the 8th Quadrennial Review of Military Compensation. This project also benefited from the comments of Major Jeffrey Stonebraker, USAF Academy; Joseph J. Cordes, Department of Economics, George Washington University; and RAND colleagues David Grissmer, Sheila Kirby, and Albert Robbert. Finally, we gratefully acknowledge the thoughtful reviews of Craig Kirkwood of Arizona State University and David Chu of RAND.

Chapter One

# BACKGROUND

## NATURE OF THE PROBLEM: A NEED FOR CHANGE

The existing officer career management system, which evolved to meet the challenges of a large military during the cold war, traces back to World War II and its aftermath. The dominant objectives then for officer management were uniformity of treatment across services and youth.[1] The Defense Officer Personnel Management Act (DOPMA) of 1980 that put today's officer management system in place continued these objectives and added high fixed promotion opportunities as a goal, which was in keeping with the civilian personnel management practices of the time. More recently, policymakers in Congress, in the Office of the Secretary of Defense, and in the military departments have expressed concerns that the different demands of the post-cold war future might require different objectives and practices for officer career management. These changed demands have been described as follows:

> Our national security strategy is evolving to reflect world changes. The Cold War strategy, dominated by the importance of containing communism, established nuclear and conventional deterrence as the primary role of our military forces. DoD emphasized aspects of military power most useful for those purposes—instantaneous readiness of nuclear bombers, combined with land- and sea-based

[1]Bernard Rostker and Harry Thie, *The Defense Officer Personnel Management Act of 1980: A Retrospective Assessment*, R-4246-FMP, Santa Monica, CA: RAND, 1993.

missile forces; large-scale, forward-deployed forces in Europe and Northeast Asia; and reinforcements ready to deploy from home.

Today's national security challenge is considerably different. There is no longer a single dominant enemy. While we are still charged with providing capabilities to fight two major regional conflicts, our attention is increasingly drawn to smaller contingencies. Instead of focusing on containment and deterrence, the National Security Strategy now emphasizes promoting democracy and economic advancement worldwide. The military component of this strategy supports creating and maintaining the stability required to allow democracy and economic growth to flourish, and staying ready to protect our interests and those of our allies and friends on short notice.[2]

## DOPMA UNDER PRESSURE

Stresses and cracks in the current officer management system became apparent during the defense drawdown and, combined with the post-cold war changes in the national strategy for our armed forces, call for a rethinking of the officer management system for the next century.[3] In particular, Congress suggested adjusting the career management system to tie it more closely to validated requirements and to achieve greater stability and longer careers. *Rethinking* is needed, because it is not clear whether a whole new system is required or whether incremental changes in the current system will suffice. The focus should fall on the *next century*, because such changes do not occur quickly, nor should they. The current officer management system went through years of discussion and debate before being enacted eight years after its introduction.[4] Any future changes to the system may entail a similarly lengthy process.

---

[2]Commission on Roles and Missions of the Armed Forces, *Directions for Defense*, Washington, DC: Government Printing Office, 1995.

[3]Congress, cognizant of the many pieces of special legislation enacted to implement the drawdown, has called for the beginning of just such a review. See Section 502 of Public Law 102-484, National Defense Authorization Act for FY 1993.

[4]DOPMA was first introduced as a legislative proposal in 1972 and was enacted in 1980. Its origins trace to the 1960 Bolte report (U.S. Department of Defense, 1960).

The DOPMA of 1980[5] was enacted in part to consolidate a patchwork of existing legislation. Congress' objectives for DOPMA were to "maintain a high-quality, numerically sufficient officer corps, provide career opportunity that would attract and retain the numbers of high-caliber officers needed, [and] provide reasonably consistent career opportunity among the services."[6] DOPMA broke new ground in establishing permanent sliding-scale grade tables, a single promotion system, and the augmentation of reserve officers into regular status, but has had difficulties controlling the significant increases and accommodating the decreases in the officer corps.

DOPMA began to show signs of stress soon after it became effective in late 1981. Although it is a good static description of the desired officer structure, it is not a flexible management tool that can respond to significant changes in officer strength and retention. During the Reagan buildup in the 1980s, DOPMA provided personnel managers many tools to increase the force, but lacked sufficient checks for controlling this growth, such as limiting the proportion of officers in the force (frequently expressed as an enlisted-to-officer ratio), or constraining the opportunity and timing of officer promotions.[7]

Because DOPMA had few tools to decrease the force, a new slate of voluntary and involuntary programs had to be developed to manage the reduction in the officer corps during the post-cold war drawdown. Recent exceptions to DOPMA and other laws Congress has enacted include new incentive programs for voluntary losses, a temporary early retirement authority, an end to the tenure protections formerly provided for officers with regular commissions, authority for increased involuntary early retirements, and grade table relief.[8]

---

[5]For more on DOPMA, see Rostker and Thie, 1993.

[6]House Report No. 96-1462.

[7]DOPMA was designed during the 1960s and 1970s when a prevailing personnel management practice was to use frequent and rapid promotion as the central mechanism for attracting and retaining young executives. In the private sector, this led to the explosion of levels in an organization in order to allow frequent promotion. In the military system, the levels (O4 to O6) were already in place; the mechanism was a "guarantee" of opportunity and timing for promotion given the fixed levels.

[8]House Reports 103-357 and 103-499 on the National Defense Authorization Acts for fiscal years 1994 and 1995 cite the U.S. Army and U.S. Marine Corps requests for grade table relief for majors and lieutenant colonels. The National Defense Authorization Act for 1996 allows the Air Force and Navy to have more O4–O6 in 1996 and 1997. The

Additional difficulties with DOPMA have arisen with the implementation of the Goldwater-Nichols Act,[9] which significantly increased emphasis on developing the joint qualifications of officers. With Goldwater-Nichols requirements superimposed upon continuing and perhaps growing demands for service-unique experience,[10] the services have had increasing difficulties managing officer careers. Too many requirements are imposed across too few career years. DOPMA contributes to this dilemma because its tenure rules constrain the length of an officer career. Most recently, concern has centered on the transition from a large force for the global conflict to a smaller one for the new international security environment.

These stresses on individual careers and in the aggregate on the services suggest the need for change. What kind of change will the future officer management system need? An important capability of a career management system is its ability to meet the future needs of the organization. Rethinking officer management objectives is imperative.

## PRIOR NDRI RESEARCH ON OFFICER CAREERS

The National Defense Research Institute (NDRI) carried out an earlier study of officer career management[11] that provided important input to the current effort. The previous study analyzed congressional and DoD concerns to determine the likely effects of different career management practices. In that study, the authors:

- Gained understanding of congressional and DoD direction for future career management.

---

National Defense Authorization Act for 1997 provides new grade tables for each service.

[9]The Goldwater-Nichols Department of Defense Reorganization Act of 1986 (Public Law 99-433) also standardized many provisions of existing law to provide more uniform basic authorities for the Secretaries of the military departments and the uniformed service chiefs.

[10]For example, the Army has been directed to provide officers to work with and gain familiarity with the reserve component.

[11]See Harry J. Thie and Roger A. Brown, *Future Career Management Systems for U.S. Military Officers*, MR-470-OSD, Santa Monica, CA: RAND, 1994.

- Constructed a general model of a career management system and analyzed the effect of its various components on officer management.
- Assessed foreign officer career management systems and changing practice of career management in public and private U.S. systems.
- Developed a range of possible future officer requirements broad enough to ensure a robust analysis.
- Designed alternative career management systems by varying the key personnel functions of accession, development, promotion, and transition.
- Evaluated the alternative systems.

## Conclusions of Previous Study

The conclusions reached in the previous study were based on a broad method of analysis designed to provide analytical information about changes that could be made in the officer career management system. We set forth alternative future systems whose designs form a "toolbox" from which needed policies can be selected to address DoD and service objectives for officer careers. We did not attempt to design or model a "best" future officer career management system. Moreover, our study questioned whether certain objectives relevant to the cold war era—e.g., uniformity of treatment across services—were correct for the future. We suggested that any effort to design a new system must start with determining and prioritizing objectives for a future officer career management system and then selecting the means—personnel functional alternatives—to accomplish them.

## Objectives Take Precedence

We further suggested that senior officials in the DoD and the military services should guide and participate in our follow-on effort, particularly to ensure that the objectives of the new management system were clearly and precisely defined. Those objectives should determine the nature of future careers for U.S. military officers.

We also proposed that the next step in rethinking officer management should be a careful review of the explicit objectives for a career management system to identify those most relevant for the future. In doing so, we need to distinguish between ends and means in officer management. In particular, we should avoid the trap of viewing some long-standing means (e.g., youth, promotion opportunity), established by DOPMA or historical practice, as ends in themselves. We do not want to preclude discovery or debate of better means to important ends.

To provide a framework for sorting out means and ends, we define several terms and then indicate how they relate to the formulation of an officer management system. In the most general sense, the primary purpose of an officer management system is to provide officers able to carry out the national military strategy (e.g., defend the nation). Objectives for any future officer management system should elaborate upon and embody the essential purposes of the system. They are selected normatively—they express what people believe the system ought to do or be like. Objectives capture important aspects of the overarching purpose, reflect lasting and desirable institutional values, or express expectations of the larger society for its military. These objectives should not be considered immutable, but subject to change only with a fundamental shift of values in the defense establishment or the society it serves.

Means contribute to achieving one or more objectives and are often selected from several competing mechanisms based on how well they achieve important ends. Their optimality can and should be subjected to regular reexamination of changing defense missions, force requirements, and environments.

Once properly identified, objectives can become the basis for selecting means. The potential pitfall, again, is that long-standing goals might be viewed as fundamental and thereby exempt from examination. For example, a youthful force is an often-cited objective of officer career management and provides much of the rationale for the tenure-limiting provisions of DOPMA and predecessor systems. But as an objective, this would seem to limit effectiveness in military skills that rely more on experience and understanding of complex systems than on the strength and stamina associated with youth. Whether such skills exist, whether they have become more dominant in military affairs, and whether individuals who have those skills should be managed differently from those in traditional skills are questions that need to be addressed. Thus, maintenance of a youthful force should be viewed as a means rather than as an objective.

An *officer management system* may be defined, in this context, as an optimal set of objectives combined with the mechanisms to implement them. DOPMA may be viewed as one such system. To reiterate an earlier point using the framework just introduced, we expect that many of the objectives underlying DOPMA will continue to be viewed as valid. However, some of DOPMA's objectives might be seen as less important and some of its mechanisms as less effective. Alternatives to them need to be considered.

## WHAT THE PREVIOUS STUDY PROVIDES TO THE CURRENT EFFORT

The previous study serves as the basis for this one. The understanding gained about future requirements, aggregation of skill groups, personnel functions, alternative career practices, and officer characteristics can apply directly. It also serves as a source of data about the effects of such practices.

### Set of Manpower Requirements for 2005

Five major determinants shape broad defense personnel requirements and the numbers of officers by service, grade, and skill required in the force: national military strategy, doctrine and operational concepts, organizational design and structures, force size and active-reserve component force mix, and technology. Clearly, the individual determinants are interdependent in their effects on officer requirements, and, for the most part, they can be viewed as external forces that the military departments and services attempt to influence but cannot unilaterally control. The last three are particularly important because of their potentially great effect on demand for officers in general and on the demand for officers with specific skills and grades in particular.

Since the beginning of World War II, rapid buildups in officer strength to meet unexpected demands for U.S. forces have been followed by substantial reductions. In its recent "Bottom-Up Review," DoD aligned the new, regionally oriented national military strategy with its expectations of future defense resources and security threats, and it concluded that in FY 1999 an active force of about 1.4 million men and women would meet the nation's security needs. For this study, we use an estimate of officer requirements associated with an active duty end strength of 1.4 million men and women, including about 175,000 officers. That is, we assume that the world evolves over the next 10 years in a straightforward way. No major detours occur, such as a significant change in the threat environment or in other determinants that would cause a large increase (or decrease) or other changes in active duty forces. However, our method is not limited to this most straightforward requirement, and we can accommodate excursions to other scenarios.[12]

## Four Skill Groups

A long-standing objective of officer career management is to satisfy officer skill requirements. In the previous study, we examined distinct skill groups that could be managed independently of each other.[13] This approach enabled us to investigate the issue of separate

---

[12]In MR-470-OSD we took into account the boom-or-bust cycle of the past to develop officer requirement options that encompassed the broad range of possibilities that might occur between the years 2000 and 2010, well beyond the current transition period. We estimated force size changes parametrically by increasing and decreasing the 1.4 million force by plus-or-minus 0.4 million. This degree of variation reasonably reflects the actual experience of the recent past. Another option streamlined the officer corps by using more civilians in positions requiring nonmilitary skills and by downgrading certain field-grade officer positions. Yet another option retained the overall active force size but examined a skill mix associated with a highly specialized officer corps. On the other hand, technology may significantly reduce the demand for officers with specialized skills. This would support the expansion of a "generalist" officer population with a broad range of operationally oriented and management skills, which we estimated using the same strength for officers. This study does not justify requirements; rather, it explores the implications of personnel management in light of a likely requirement, which was the one accepted by the executive and legislative branches at the time of the study.

[13]The titles used to name and describe the skill groupings in our construct in some cases have already well-established meanings with a lengthy history or cultural acceptance within the military. The primary purpose of our use of these terms is to define distinct sets of officer requirements in large aggregate groupings with skill characteristics common to all services that suggest the need for separate career management activities. For example, we intend that the term "line" classify one set of unique mili-

career management systems for distinct skill groups. If skill groups are managed separately rather than uniformly in careers within and among the services, one might expect to have uniformity in careers within skill groups with overall service careers different insofar as service skill composition differs.

We identify four separate skill groups: line, specialist, support, and professional. We define *line* skills as uniquely military, particularly those directly involved in combat operations and related military functions (e.g., tactical operations officers, intelligence officers). The *specialist* group requires military skills and recurring assignments that rely on advanced education or high cost, long-duration training or experience (e.g., aviation maintenance officer). *Support* group skills are analogous to civilian white-collar occupations that support the functioning of military organizations where general military experience is desired or assist task performance (e.g., logistics, transportation, comptroller). *Professional* skills are civilian professional skills not usually requiring significant military experience (e.g., medical, dental, legal, and chaplain).[14]

The amount of desired military experience also differs by skill group. Line skills require military experience; specialist skills need both military experience and technological expertise; support skills need experience in those skills tempered by adequate military experience; and the professions require only limited military experience to complement professional knowledge.

Most skills in the Army, Air Force, and Marines Corps (fewer in the Navy) are now included in the line category for competitive man-

---

tary skills generally acquired through established military education, training, and experience. We recognize that this use of the term has a meaning different from its historical antecedent. We do not include the more common notion of line as the only group of officers who can exercise command. The latter usage has lengthy history in the military services, especially in the sea services, but is not intended here to be restrictive. Positions that include the exercise of command may be in any skill grouping, and our line would certainly include any officer requirements of that nature not elsewhere considered. For instance, the position of commander for a specialized organization that required the lengthy education and training of a "specialist" would be in the specialist skill grouping. Adherence to our explicit definitions and usage is essential to prevent confusion.

[14]Our primary objective was to demonstrate, at an aggregate level, officer requirements that are largely homogeneous in nature for separate management policies. There is no intention to imply a high degree of precision to this regime.

agement, even though some skill groups have traditionally achieved greater promotions and higher positions. However, if specialist and support officers were as apt as line officers to achieve the highest positions and were considered central to the profession, then there would be no need to manage them separately. Certainly the present system manages two skill groups—line and professionals—in fundamentally different ways, and there is conceptually no reason that this cannot be extended to more than two skill groups.

We are not asserting that certain skills should be managed differently from others, but we will examine how such skill groups could be managed differently, should that approach meet management objectives.

## Four Major Personnel Functions

In building alternative career management systems, we used variations available to policymakers in the design features of four personnel functions—accessing, developing, promoting, and transitioning. These personnel functions integrate the individual's capabilities with the requirements of the position and affect outcomes. Manipulating personnel functions can provide variation within a career system depending on the choices made about the system's various aspects. The aspects we consider here emerged in the prior study as central to the design of a management system. We gleaned them from our analysis of the U.S. military, foreign military establishments, public and private sector systems, and the career management literature. Our methodology allows for additional aspects to be evaluated. We will develop a career management system around choices for various aspects of these four personnel functions.

**Accessing** deals with choices about entry into careers and includes policies pertaining to such matters as initial active duty commitments, use of different sources for acculturating prospective officers [e.g., academies, Reserve Office Training Corps (ROTC), Officer Candidate School/Officers' Training School (OCS/OTS), etc.], the point in a career profile at which new entrants might enter into careers, and the standards (e.g., education, physical, aptitude, moral) officers must meet at entry.

**Developing** concerns the patterns of education and assignments that define careers. Policies include the length and number of assignments in an officer's career; the nature of those assignments, usually represented by a career path; and the education (both military and civilian) needed.

**Promoting** deals with movement to higher levels of responsibility or the failure to do so. Policies cover opportunity (probability of promotion to each grade), timing (expected time in service before promotion to particular grades or the time in each grade between promotions), and the basis for promotion (e.g., emphasis on merit or seniority).

**Transitioning** concerns how and when to end careers. Policies cover tenure (limits on involuntary separation that protect the individual), minimum service required before payment of an immediate retirement annuity, the amount of service required before an officer is vested for some future retirement annuity, and the maximum service allowed prior to mandatory retirement.

## Characteristics of an Officer

In this study, we are most interested in a subset of the nation's human capital—those who have the ability and motivation to become officers. In the research for the earlier study, we established that the military officer is expected to be of "high quality," by which we mean leaders who are intellectually and physically vigorous.[15] Officers are also expected to be conscientious (willingly expend effort toward goals) and versatile (can accomplish multiple tasks now and can learn as the environment changes). Moreover, they continue to be members of a profession, in this case the military, which means they wear uniforms and may be shot at; must be knowledgeable about a rigorous body of military science; and adhere to values such as integrity. We assume in this study that excursions in career management will not change the nature of the officer corps.

---

[15]A RAND researcher who assisted in the previous study, Paul Bracken, draws a useful distinction between leading and managing. Leadership deals with overall direction of the organization, the need for change, and a sense of the possible. Management deals with complexity and with attending to the many interacting factors needed for successful implementation.

## Data and Analysis

We collected and used a substantial amount of data in analyses for the previous study. We draw on both the data and the analyses for this study, particularly for the quantitative information. For example, we gathered information on the cost of acquiring an officer through various accession programs. We use those costs here to study the effect of policy choices about accessing. In other instances, we draw on analyses of the earlier and other studies that were based on quantitative information. For example, NDRI has modeled the effect of various retirement programs on the flow of personnel into and out of the system. We have used that modeling to inform our judgments about the policy effect of vesting alternatives. We do not create new sources of data in this report, but instead use documented data and analyses to measure the effects of alternatives against objectives.

Not all the detailed information needed to assess alternative career management practices is available. For example, we have data about what it costs to acquire an officer, but we do not have estimates about how those costs might change under every alternative. Also, the qualitative objectives (e.g., provide career satisfaction), have not been analyzed or measured. Where information was unavailable, we assessed the likely direction of effect. For example, in considering compatibility with civilian careers, it is clear that having no provision for a pension would be completely incompatible. Thus, we could easily determine the likely direction of the effect.

In summary, our previous study and other research provide

- assumptions,
- aspects of and alternatives for personnel functions that define a career system, and
- data about and analyses of such aspects.

This study

- specifies objectives for career management and
- uses the weighted objectives to determine preferred alternatives.

Chapter Two

# PURPOSE OF THIS STUDY, METHODOLOGY, AND ORGANIZATION

This study's three goals are: to determine objectives for a future officer career management system (OCMS), design an OCMS that reflects DoD decisionmakers' preferences for these objectives, and assist decisionmakers by providing insights on addressing change. Figure 2.1 depicts the process the study followed to accomplish those goals. The shaded boxes connote research team activities, and the unshaded box indicates input from the policymaker group.

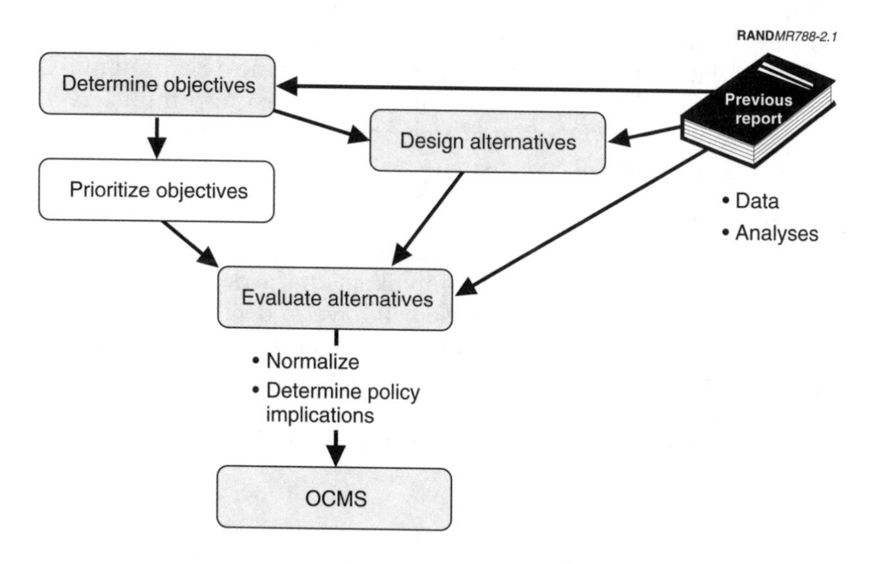


**Figure 2.1—Study Process**

As the diagram indicates, the data and analyses of the earlier research (MR-470-OSD) inform the entire process. In particular, the prior study has a wealth of information on the ability of alternative designs for personnel management functions to achieve objectives. That information assisted us as we designed the alternatives, formulated the objectives, and, most especially, when we evaluated them.

To arrive at an OCMS, we apply multiobjective decision analysis methodology, which is a way of quantifying the preferences of decisionmakers. We determine the objectives of an OCMS and the relative preferences of policymakers for those objectives (the left side of Figure 2.1). We translate the preferences of policymakers for different objectives into weights that represent the collective consensus of a group of high-level DoD personnel policymakers for each objective.[1] We also design alternatives for accomplishing the objectives (right side of the diagram—note that the process of determining objectives influenced our design of alternatives). Applying the multi objective decision analysis methodology enables us to evaluate the alternatives the research team designed in terms of how well each alternative accomplishes each objective and then how well it accomplishes the collective set of objectives. As the diagram implies, the design of the alternatives occurs in parallel with the determination of the objectives. The set of alternatives that best accomplishes all objectives becomes the officer career management system, if the alternatives are internally consistent.

This report includes explicit objectives for a future officer management system and evaluates personnel function alternatives that best achieve the preferences of decisionmakers for the determined objectives. The Department of Defense will have to decide about implementing such a system. The career management system proposed here suggests promising future directions, many of which differ from those of today. Implementation of these new directions, if warranted, requires commitment to the objectives that prompted their

[1]The Assistant Secretary of Defense (Force Management Policy) declared the services' input crucial to this effort and chartered two groups: a working group of service representatives at the O4–O6 officer/civilian-equivalent level, and a policymaking group of senior military and civilian leadership. These groups are described later in the report.

selection. It will also require additional analysis and possibly even some testing of alternatives prior to full implementation.

## HOW THE REPORT IS ORGANIZED

The organization of this report follows the steps of the methodology we applied to determine the officer career management system. It has five remaining chapters and six appendices.

**Chapter Three** presents the elements of the officer career management system and defines the specific alternatives within it that require decisions.

**Chapter Four** describes the process we used to determine the objectives of the career management system and the relative weights accorded to them by policymakers.

**Chapter Five** illustrates how the methodology is used to make decisions about the various alternatives in the career management system by giving a generic example.

**Chapter Six** presents the officer career management system that results for line officers from the set of weighted objectives. A sensitivity analysis shows the effect of changing the weights on objectives.

**Chapter Seven** offers concluding observations.

The six appendices contain:

- The interview questionnaire—the vehicle we used to determine policymaker preferences for objectives.
- An example of one aspect's determination. Appendix B shows the complete process for evaluating one set of alternatives to decide which entry point for line officers best meets policymaker preferences.
- The value functions. Appendix C contains the quantitative values we used to determine the policy implications of various alternatives.

- The data for the sensitivity analysis. Appendix D contains the data we used to determine the sensitivity of alternatives to different objective weights.
- Feasibility of supply. Appendix E describes our modeling to ensure that the career management system derived from the objectives-based methodology could supply the requisite numbers and grades of officers needed.
- OCMS for Other Skill Groups. Appendix F shows the resulting career management system for other skill groups of officers—specialist, support, and professional—based on the research team's weighting of objectives.

Chapter Three

# PERSONNEL FUNCTIONS, ASPECTS, AND ALTERNATIVES

## BACKGROUND

Our earlier officer management study (MR-470-OSD) indicated that the officer career management system is composed of the four personnel functions described in Chapter One. Each function has a number of aspects that define the specific characteristics of an OCMS. For example, the aspects of accessing evaluated in this study are entry point, initial tenure, the type of pre-entry acculturation recommended, and each officer's amount of obligated service. Each aspect could take a variety of forms, or alternatives. For instance, for entry point, an aspect of accessing, we consider four alternatives: all officers enter at the earliest career level, transfer laterally from civilian life at any point in a career, enter laterally from prior service—commissioned or enlisted—or the reserve component, or some combination. Designing an OCMS entails making decisions about these alternatives, which in turn determines the aspects.

The research team designed the alternatives considered here. As Figure 2.1 shows, we drew heavily on the earlier report. It was clear from that research that some alternatives would figure into any officer management system—for example, only certain things can happen to an officer not selected for promotion: separation, retention in grade, and so forth. For these, we wanted to ensure that we included a range of alternatives bounding the status quo that was robust enough to capture most feasible options. In other cases, the previous research showed that some ideas were particularly attractive, such as longer careers. In these cases, we did not necessarily bound the status quo; rather, we simply ensured a wide range of feasible options. And in yet other cases, the objectives for the career management system, which we were developing in parallel with the alternatives, suggested additional possibilities.

Table 3.1 shows the four functions, the aspects of those functions that require decisions, and the alternatives we designed for this study.

**Table 3.1**

**Functions, Aspects, and Alternatives Considered**

| Function | Aspects | Alternatives |
| --- | --- | --- |
| Accessing | Entry point | Lateral from civilian<br>Lateral from military<br>Year 0 |
| | Initial tenure | 2 years<br>4 years<br>6 years |
| | Pre-entry acculturation | None<br>Educational, high-intensity, short<br>Educational, low-intensity, long<br>Educational, high-intensity, long<br>Experiential, medium intensity, medium |
| | Amount of obligated service for education, training | 0.5 year<br>1 year<br>1.5 years<br>2 years |
| Developing | Career selection point | None<br>5–10 years<br>8–10 years<br>> 10 years |
| | Effect of nonselection | Separation<br>Migration to new skill |
| | Average assignment length | Decrease by two-thirds of average<br>Decrease by one-third of average<br>Current length<br>Increase by one-third of average<br>Increase by two-thirds of average |
| | Military and civilian education | Current amount<br>2 years more<br>2 years less |

**Table 3.1—continued**

| Function | Aspects | Alternatives |
|---|---|---|
| Promoting | Promotion zone | Time in service<br>Time in grade<br>Combination |
| | Length of zone | Narrow (1–2 years)<br>Broad (3–8 years)<br>Open |
| | Opportunity | Fixed<br>Selective (based on requirements) |
| | Nature of continuation | Guaranteed<br>Based on requirements |
| Transitioning | Vesting point | 4–9 years<br>10–15 years<br>20 years |
| | Transitional ability of the system | Tenure<br>Voluntary separation incentives<br>Neither tenure nor incentives |
| | Maximum career length | 30 years<br>35 years<br>40 years |
| | Separation rates in first 10 years | High<br>Medium<br>Low |
| | Retirement annuity point | 15 years<br>20 years<br>25 years<br>30 years<br>35 years<br>40 years |

## DECISIONS ABOUT ASPECTS OF PERSONNEL FUNCTIONS DEFINE THE OCMS

We next detail each of the four personnel functions and describe the alternatives designed for each.

### Accessing

There are four aspects of accessing considered for the future OCMS.

**Entry Point.** Entry point refers to when and in what way an individual can become an officer. There are three alternatives. The first alternative, lateral entry from civilian life, permits an individual to enter as an officer, receive career credit for years of experience and education in the private sector, and apply relevant civilian skills to a military career. For example, the military currently offers lateral entry opportunities to physicians. The second alternative, lateral entry from the reserves or with prior active military service, grants the individual credit for experience gained in the reserve component or in prior active duty service. Prior active service could be previous enlisted service. The third alternative, year 0, represents entry at the beginning of a career path. This third alternative is the current norm for line officers.

**Initial Tenure.** Initial tenure pertains to the number of years an officer has to demonstrate potential for continued service. It is in effect a probationary period that recognizes that some individuals learn and adapt more slowly than others or have more or less opportunity to demonstrate proficiency. Promotion to O-3, which nominally occurs at four years, is the first significant officer screening in the current system. For this analytical exercise, we bound the current practice with two alternatives. Thus, the three alternatives evaluated in this study are two, four, and six years of initial tenure.

**Pre-Entry Acculturation.** Pre-entry acculturation reflects the duration and intensity of the process by which a new officer has been exposed to the military culture before entry. This aspect of accession has five alternatives. Under the first, officers enter the career system without any prior acculturation. The second alternative is educational in nature, of high intensity and short duration, and resembles an OCS-like acculturation process. The third alternative also is an educational experience, but of low intensity and long duration. It resembles an ROTC-like experience, which takes two to four years but does not consume every day of that period. The fourth alternative is an educational experience of long duration, but is of high intensity. It might resemble the current academy experience. Finally, the last acculturation process alternative is an experiential, rather than educational, process. It is of medium intensity and medium duration and is designed to represent prior enlisted service considered as acculturation for future officers.

**Amount of Obligated Service.** This decision applies to the time that officers should serve on active duty to recoup a training, education, or experience investment made before or immediately after entry. The alternatives include half a year, one year, one and a half years, or two years for each year of investment benefit. For example, if an incoming officer received four years of scholarship benefits, the four alternatives would indicate obligated service of two, four, six, or eight years.

## Developing

This effort considers four aspects of officer development: the point at which an officer is selected for career status, the effect of nonselection for career status, the average assignment length, and the best amount of military and civilian education.

**Career Selection Point.** For many, entry as an officer carries with it an expectation of continuing into a career track (augmenting)[1] given successful performance in the entry position and available future positions in which to serve. However, because the line skill group typically has more junior officers than other skill groups, not all officers who want to will be able to serve full careers as line officers. Fewer senior positions are available in the line specialty than others. The career selection point is the time in an officer's prospective career when individual potential is matched with needs of the service for longer-serving officers in higher grades. In other words, selection for career status generates expectations for additional service as long as performance is satisfactory and the officer's skills and experience are needed. Four career selection point alternatives are taken into account: none, which means that no separate decision is made about continuing (i.e., all who wish to continue may); between five and ten years, which represents a notional early- to mid-career selection point (the status quo is between eight and ten years of service, the period during which officers are promoted to O4 and, under DOPMA's provisions, gain tenure until 20 years); after ten years,

---

[1]Augmentation, the process of becoming a career officer, begins after initial entry into military service and ends with judgments about officer potential to serve for a full career or until separated.

which represents a notional mid- to later-career selection; and after 15 years, a notional late-career selection point.

**Effect of Nonselection for Career Status.** There needs to be a management policy determining the effect of nonselection for career status. The two alternatives in this study are separation or directed migration to another skill. Thus, a line officer not selected for career status would, under the first alternative, leave the military. Under the second alternative, the officer might continue in another skill group, perhaps as a specialist or support officer.

**Average Assignment Length.** Maintaining the current average assignment length, lengthening the average by either one-third or two-thirds of an average assignment, or shortening it by the same amounts are the five alternatives for average assignment length. Thinking about changes to the current system requires some definitional groundwork regarding what constitutes a single assignment. If a change in jobs places an officer in a new billet that is substantively similar to his previous billet and does not require a location change, we regard that as a single assignment. A change in location or substantive change in daily duties qualifies as a new assignment.

**Amount of Military and Civilian Education.** This study evaluates changes in the amount of service-provided military and civilian education offered to officers during their careers. The two kinds of education are evaluated independently. The alternatives are two years more than the current amount over the length of an average career, the current average amount of education, or two years less over the length of an average career.

## Promoting

This study considers four aspects of the officer promotion function: whether time in grade or time in service determines the promotion zone; the length of the promotion zone, whether the promotion opportunity is fixed or selective, and whether a fixed or variable number of officers continue after nonselection.

**Promotion Zone.** This study takes into account three alternative ways of determining the promotion zone. First, the time in service determines when an officer is considered for promotion. Second, the time in a particular grade determines when an officer is eligible for promotion (this represents the current promotion system). A time-in-grade system can move officers through the grades much more quickly than the time-in-service system. Third, a combination of the two alternatives is considered. In this alternative, time-in-grade given a minimum requirement of time-in-service would determine when an officer is eligible for promotion.

**Length of Promotion Zone, Nature of Promotion Opportunity, and Nature of Continuation.** The final three aspects of promoting officers are considered together because of their normal interrelationship in the promotion function. The length of the promotion zone determines the number of years an officer is considered for promotion. "Above the zone" and "below the zone" distinctions disappear. The alternatives considered are a narrow zone of one to two years, a broad one of three to eight years, or an open zone. In a system with an open zone, officers remain eligible for promotion until they leave active duty.

Promotion opportunities can be either fixed or selective. A fixed system guarantees the percentage of officers who will be promoted. This rate is determined years in advance of the promotions and is "guaranteed" to the officers. The current officer promotion system is a fixed system. A selective promotion opportunity promotes officers to satisfy grade requirements. It is worth noting that the fixed promotion rates could be consistently lower than the selective rates; the difference is not in the number of officers promoted, but in the way the promotion opportunity is determined.

Similarly, a system with fixed continuation can guarantee the percentage of officers who can remain if not promoted far in advance of the promotion cycle, or it could be based upon the requirements for officers in skill and grade under a selective continuation system.

These three aspects—the length of the zone, the nature of promotion opportunity, and the nature of continuation—are combined into a single decision problem with ten possibilities to consider. Thus, one possibility is a narrow promotion zone with a fixed promotion opportunity and selective continuation. The other possible combinations are the various permutations of each of the three aspects, with two exceptions. Open promotion zones using either fixed or selective promotion systems are only compatible with a fixed continuation rate of 100 percent. Because the officer never leaves the promotion zone, selective continuation has no point. Table 3.2 arrays the possible options.

## Transitioning

Five aspects of transitioning—the policies pertaining to leaving the officer career system—are contemplated. Vesting and the ability of the system to shrink itself are evaluated independently. The other three—maximum career length, rate of separations in the first ten years of service, and retirement annuity point—are considered in combination.

**Vesting Point.** A vesting policy is established when an officer becomes eligible for some later retirement annuity. The amount of the annuity depends upon the amount of time served before leaving the service. Three alternatives are considered. Vesting at between four and nine years of service can be considered an early career vesting point. An alternative vesting point between ten and 15 years represents a mid-career vesting policy. Finally, the current practice, vesting at 20 years, is the third alternative.

**Transitional Ability of the System.** The transitional ability of the officer career management system represents the flexibility of the sys-

**Table 3.2**

**Possible Combinations for Promotion**

| Alternative | Length of Zone (years) | Opportunity | Nature of Continuation |
|---|---|---|---|
| 1 | 1–2 | Fixed | Selective |
| 2 | 1–2 | Fixed | Fixed |
| 3 | 1–2 | Selective | Selective |
| 4 | 1–2 | Selective | Fixed |
| 5 | 3–8 | Fixed | Selective |
| 6 | 3–8 | Fixed | Fixed |
| 7 | 3–8 | Selective | Selective |
| 8 | 3–8 | Selective | Fixed |
| 9 | Open | Fixed | Fixed |
| 10 | Open | Selective | Fixed |

tem to decrease in size or skill composition quickly by separating people. There are four alternatives. First, we consider tenure, which grants continued service to individuals who reach certain plateaus, or grades, and thus protects them from separation for established periods of time. Second, we include a practice that provides incentives for voluntary separation. In essence, this system mitigates any disadvantages of not promising tenure by providing financial inducements when separation is needed. Third, an alternative system offers neither tenure nor incentives to separate. Fourth, we evaluate a practice that offers separation pay to those officers involuntarily separated from service.

**Maximum Career Length, Rate of Separations in First Ten Years, and Retirement Annuity Point.** The final three aspects of transitioning are so closely related to one another that they require a decision in concert. The alternative maximum career lengths considered are 30, 35, and 40 years. The rates of separations in the first ten years were set at either high, medium, or low. The retirement point—when a retired officer receives a retirement annuity—included alternatives of 15, 20, 25, 30, 35, and 40 years. The resulting decision problem had 54 possible alternatives.

In summary, the aspects of the four personnel functions *(accessing, developing, promoting, and transitioning)* define an officer career management system. Which alternatives are best depends on the objectives of that system, the subject of Chapter Four.

Chapter Four

# OBJECTIVES OF THE OFFICER CAREER MANAGEMENT SYSTEM

Chapter Three discussed the alternatives we designed for the career system. Choice of alternatives hinges on what the officer career management system is trying to accomplish, that is, its objectives. This chapter describes how we identified the objectives for an officer career management system and how we determined policymaker preferences for them.

Determining the objectives that will be the basis for decisions is a crucial step in multiobjective decision analysis, because "the set of objectives used to evaluate the alternatives . . . is the foundation on which any analysis rests."[1]

Formulating objectives has several important effects. First, specifying objectives compels decisionmakers to recognize and define their goals. Second, it fosters ideas and creates alternatives.[2] Third, evaluating alternatives and selecting the best combination for a future career system hinges on the set of objectives. Finally, the objectives permit discussion, description, and debate about the proposed system. E. S. Quade notes that one way policy analysis and systems analysis, which include multiobjective decision analysis, differ from economics and other fields is in "consider[ing] the problem of what ought to be done as well as how to do it."[3] A well-defined set of objectives permits this consideration.

---

[1]Ralph L. Keeney, "Structuring Objectives for Problems of Public Interest," *Operations Research*, Vol. 36, No. 3, May–June 1988, p. 396.

[2]Ibid., p. 404.

[3]E. S. Quade, *Analysis for Public Decisions*, New York: American Elsevier, 1975, p. 84.

We cast a wide net in our effort to capture objectives and based our effort on multiple sources of information, including an extensive review of career management literature; intensive discussions with experts; and multiple seminars with military and nonmilitary influencers, stakeholders, and decisionmakers over a three-year period. We convened two major discussion and brainstorming sessions to determine objectives. One was attended by retired four-star generals and admirals, other general and flag officers, and senior executives with backgrounds in the four services, the Office of the Secretary of Defense (OSD), Congress, and academe. These individuals participated in a rich discussion of goals, assumptions, and objectives for officer management, based upon their career and lifetime experience. The second discussion seminar included area experts and analysts from the military manpower and personnel community. As the study proceeded, the working group and the policymaker group of service members and DoD civilians participating in our effort reviewed potential objectives. Internal discussion with RAND colleagues also helped shape the set of objectives.

Identifying the correct objectives is a vital aspect of multiobjective decision analysis. Brainstorming discussion sessions with military manpower experts were an important element of identifying objectives, but the task of winnowing the discussion results into a succinct but complete set of appropriate objectives fell to the authors. Many of the objectives initially suggested do not appear in the final set.

Throughout the process of identifying objectives, we encountered problems typical of such efforts. Objectives may not be clearly envisioned. Multiple objectives may conflict with one another, and objectives are frequently difficult to distinguish from means to accomplish them. For example, total quality management or rapid promotion may be proffered as objectives. However, each actually pursues another goal. Total quality management improves product reliability, which in turn yields increased customer satisfaction. In this case, customer satisfaction is the objective and total quality management a means to satisfy it. In the military, rapid promotion is cited as a way to ensure career satisfaction and to retain officers. But retention is pursued for the sake of meeting the military's requirement for experienced officers at multiple levels of leadership responsibility. Thus, rapid promotion is actually a means to accomplish three potential objectives: provide career satisfaction, satisfy skill requirements, and meet grade requirements.

To help us resolve these problems, we used a hierarchical approach, which enabled us to explore the many concerns voiced during our research. We considered why each concern was important. Often the question "why" produced additional concerns or placed the original concern within the growing hierarchy. For example, a concern mentioned above and frequently raised in the discussion groups was that officers should be promoted rapidly. Asking "why?" resulted in an objective to provide career satisfaction. Again, when we considered "why" career satisfaction was important, it became apparent that it was worthwhile to foster careers. Why is it important to foster careers? Because it is an important goal in and of itself. As discussed in Appendix B in the previous study, the term "career" refers to a long-term series of related positions in the profession of "officership." This view of an officer's career is important because it promotes the commitment and dedication required to develop a highly competent officer corps. With its focus on the profession, our definition contrasts with others that focus on the individual and define a career as a sequence of jobs, regardless of profession or level. Our definition implies a path of career movement to include a clear pattern of systemic advancement.[4]

Determining that an element is a goal in and of itself means that the top of the hierarchy has been reached. In the descriptions of the resulting objectives, foster careers is one of four broad considerations, and providing career satisfaction is one of 11 objectives. Promotion practices appear later in the process as a means of obtaining the objective of career satisfaction.

Once a rough set of objectives was hierarchically arrayed, we compared the set with guidelines for methodologically valid objectives. In multiobjective decision analysis, objectives must meet four criteria to be methodologically valid.[5] They must be complete, not re-

---

[4]Douglass T. Hall, *Careers in Organizations,* Pacific Palisades, CA: Goodyear Publishing Company, Inc., 1976.

[5]See Craig W. Kirkwood, *Multiobjective Decision Analysis,* Department of Decision and Information Analysis, Arizona State University, 1995. (Forthcoming as *Strategic Decision Making,* Belmont, CA: Duxbury Press.)

dundant, operable, and the set must be of a reasonable size (i.e., cannot be too numerous or too few). In other words, the objectives must represent a complete list and adequately cover all concerns relevant to the effort; additional objectives cannot be determined later in the effort. The objectives must not be redundant or depend upon one another; that is, an objective's rank should not depend upon the ranking another objective receives. To be operable, the objectives must be tied to alternatives to evaluate and assess them. Finally, the set of objectives must be sufficiently small to be understandable and explicable.

We employed three techniques to reduce the many objectives suggested into a manageable set. First, some objectives were combined. For example, we combined the proposed objective "maintain balanced societal representation in the force" into "provide a high opportunity to serve." When we conducted the interviews to gather priorities for the objectives, we presented this objective as "a high and representative opportunity to serve" and explained it as a combination of the two issues. Likewise, "provide career satisfaction" is defined as work that is challenging and responsible, adequate compensation, including promotion, and satisfaction of family needs. Initially, these were three separate proposed objectives.

Second, other proposed objectives were split into separate objectives. "Meet active skill needs," "meet active grade needs," and "meet active experience needs" initially were proposed as a single objective. We determined them to be separate and vital issues that required individual consideration.

A third technique was to incorporate suggested objectives into our assumptions. For example, both "attract highly able and conscientious entrants" and "separate the unqualified" appeared on early objective lists. However, we assumed that the officer corps was composed of high-quality individuals and that any changes to the management system would not affect the officer quality level. Likewise, we assumed that those who performed poorly would be removed from the system. However, we did not evaluate the effect of changes on supply, and these would have to be assessed before making any substantial changes.

## THE RESULTING OBJECTIVES FRAMEWORK

At the top of the hierarchical framework of objectives is the overarching purpose of officer career management: to develop and deploy the nation's human capital to satisfy national security needs. The management system strives to match human performance and satisfaction to the missions of military organizations. A subset of the nation's human capital is particularly relevant to this system—those with the ability and motivation to be the high-quality officers previously defined.

Under the overarching purpose of the career management system, four broad considerations organize a total of 11 objectives, as shown in Table 4.1.

The OCMS must develop and manage officers to satisfy the national security needs. Thus, the first broad consideration is to meet military manpower requirements. This is generally stated as having the right mix of experience, skill, and grade to meet the needs of the active force and to provide reasonably junior prior active service officers to meet reserve component needs. Thus, these are the four objectives organized under the broad consideration of meeting military manpower requirements.

**Table 4.1**

**Broad Considerations and Objectives**

| Broad Consideration | Objective |
|---|---|
| Meet military manpower requirements | Meet active experience needs |
| | Meet active skill needs |
| | Meet active grade needs |
| | Meet reserve needs |
| Be consistent with public values | Keep costs reasonable |
| | Provide high opportunity to serve |
| Maintain military culture | Emphasize cadre with military culture |
| | Inculcate culture prior to or at entry |
| Foster military careers | Provide career satisfaction |
| | Provide career opportunity |
| | Be compatible with civilian careers |

The second broad consideration addresses the need for the military to be consistent with public values about military service. This consideration includes two objectives that reflect what policymakers believe the public cares most strongly about: "Keep costs reasonable" and "provide high and representative opportunity to serve" the nation as officers.

Third, the participants in the brainstorming sessions and seminars repeatedly emphasized the importance of military culture. For this study, we defined culture to include the values, attitudes, and beliefs held by people in the organization. The two objectives under this consideration reflect two approaches to preserving military culture. The first approach emphasizes the military cadre who have committed themselves to the culture, embody it, and transmit it to others by example and by acting as mentors. The second approach stresses transmitting the military culture and inculcating its values in individuals before or at their entry as officers.

The fourth broad consideration for officer career management is to foster careers. This consideration rests upon an assumption that the military officer system should remain a career system, rather than a system of individual short- or long-term jobs. Given this assumption, three objectives are included under this broad consideration. The first is to provide career satisfaction to the officers in the system. This study defines career satisfaction as having challenging and responsible work, having adequate compensation and promotion opportunity, and accommodating family responsibilities. The second objective under fostering careers is to provide career opportunity. Career opportunity is defined as being able to serve in a career for a reasonably long period of time. The final objective under this consideration is compatibility with civilian careers. Compatibility with civilian organization career management practice in the United States is important because the military is a part of society and should not be divorced from broad societal trends. Thus, military officer careers should have some similarity with their civilian counterparts.

## DETERMINING PREFERENCES

We note that the determination of these 11 objectives does not imply any equivalent importance. In other words, by defining the objectives to include both "emphasize cadre with military culture" and "be compatible with civilian careers," the project team is not declaring that those two, or any of the other objectives, have equal importance. Instead, the next step is to determine the relative importance of these objectives to the decisionmaking group who would design and implement the next OCMS.[6]

The Assistant Secretary of Defense (Force Management Policy) designated a policymaking group of military and civilian executive decisionmakers to consider the next OCMS. In addition, a working group of military and civilian staff experts in the Pentagon was designated to assist.[7] These groups were key to evaluating and prioritizing the 11 objectives for the future officer career management system.

The policy group was comprised of three senior OSD officials, a senior executive from each military department, and a senior officer from each military service. It included five assistant or deputy assistant secretaries and three three-star officers. They occupied the following positions:

- Assistant Secretary of Defense (Force Management Policy)
- Principal Deputy Assistant Secretary of Defense (Force Management Policy)
- Deputy Assistant Secretary of Defense (Military Personnel Policy)
- Assistant Secretary of the Navy (Manpower and Reserve Affairs)
- Deputy Assistant Secretary of the Air Force (Military Personnel)
- Deputy for Military Personnel Policy, Office of the Assistant Secretary of the Army (Manpower and Reserve Affairs)
- Deputy Chief of Naval Operations (Manpower and Personnel)
- Deputy Chief of Staff (Manpower and Reserve Affairs) HQ, USMC

---

[6]This group is not a representative sample of the population, nor was it intended to be.

[7]The 8th Quadrennial Review of Military Compensation (QRMC) also provided input and assistance and shared many of its findings.

- Director of Professional Development, HQ, USAF
- Director of Officer Personnel, HQ, USA.

The working group contained 22 personnel policy experts or personnel policy analysts from OSD and all the military departments and services at the grades of O4, O5, and O6 or civilian equivalents. The two groups provided 30 people well experienced with military personnel and currently in positions that deal with the key officer personnel issues of the services and the DoD.

There are several ways to assign priorities to a given list of objectives. One method simply asks participants to divide 100 percent among all objectives, assigning the greatest weight to the ones they regard as the most important. Total points awarded cannot exceed 100. We chose instead to ask our decisionmakers to choose between paired choices of both the broad considerations and the 11 objectives. Additionally, we asked about preferences for the various personnel functions and the management principles of uniformity and flexibility, both across services and skill groups. These last two sets of questions and their use will be explained in more detail later. The questions were posed in a structured interview; see the questionnaire in Appendix A.

The structured interview included 38 paired questions that asked group members what they thought best achieved either a purpose or an objective for the year 2005 in an environment that had evolved with no major departures from today. Thus, we were striving to achieve an external fit with the most likely future environment. The same process could be applied to determining preferences for other scenarios.[8] The objective ratings were not awarded based on their sensitivity to policy alternatives.

---

[8]The participants were first asked to choose between paired objectives subordinate to a broad consideration. For example, one question was "For fostering careers, do you consider providing career satisfaction or providing career opportunity more important?" The interviewee chose one of the two alternatives and then weighted the choice from 1 to 9 based on the strength of the preference. A weight of 1 signified that the two alternatives were equally important to fostering careers. A weight of 9 indicated an "extremely strong" preference for one objective relative to the other. All interviewees were provided the same explanation of the meaning of the questions and the weights. Once the interviewee had completed the choices among the paired objectives, he was asked for preferences among the four broad considerations.

The interviews with the senior decisionmakers and with the working group members were conducted individually, which enabled us to resolve any confusion over definitions or inconsistency. We are comfortable that semantic issues were handled well enough to yield reliable interview results.[9]

The paired structure of assigning preferences offers several advantages. First, paired choices provide a structured means of discriminating among multiple objectives. The participants had a relatively easy time choosing between two objectives, whereas choice among multiple objectives or a ranking exercise would have appeared more formidable. Second, among the objectives on a longer list, often one or two appear that are commonly perceived as the most important. Paired choices compel the interviewees to make multiple decisions reflecting their views, and their individual preferences emerge. In addition, it is less likely that interviewees will offer what they perceive as the most acceptable answer.

After completing the structured interviews for both groups, we conducted a feedback session with the working group.[10] During this session, we showed each individual the rankings we computed from his choices among the four broad considerations and the 11 objectives. Each person also received an anonymous account of the other group members' responses. At the conclusion of the feedback sessions with the working group, we collected the written feedback and

[9]The other key problem that can arise from a paired interview method is inconsistency, and there are two kinds: qualitative and quantitative. Qualitative inconsistencies result when an individual prefers Option A to Option B, and Option B to Option C, but Option C to Option A. We did not observe qualitative inconsistencies in our interviews. Quantitative inconsistencies result when an individual prefers Option A twice as much as Option B and Option B five times as much as Option C. Theoretically, to be consistent, this individual should prefer Option A ten times as much as Option C. Because of the limitations of the software we used to convert the interview results into overall preferences, we limited the interviewees to a nine-point scale. (We used the software *Expert Choice* to assist us in this process.) Thus, some quantitative inconsistencies were unavoidable when respondents felt strongly about preferences. However, these quantitative inconsistencies were minimal and did not affect the outcomes.

[10]We also conducted structured interviews with the members of the 8th QRMC and presented their responses back to them during a feedback session. While participating with the QRMC was not part of the research proposal, it was determined to be in the mutual interest of our study as well as the QRMC. This permitted us an additional opportunity to test our interview and feedback process.

invited the participants to change their preferences if they felt their views had been misrepresented. Four out of 22 individuals altered their paired choices once they saw the preferences that resulted from their initial responses. However, these changes were minor and affected only a couple of the choices they had been offered. In addition, only one person changed the order of his priorities and then only on a single paired choice. The other changes kept the same prioritized order but rounded the prioritized values slightly. The majority of the group left their preferences unchanged. Thus, the feedback session tended to confirm the structured interview and calculation of individual preferences, and we therefore did not convene the policymaking group to review their responses.

## OBJECTIVE WEIGHTS

The preferences provided by each of the policymaking group members serve as the basis for determining the future officer career management system. There were individual differences among these preferences for the 11 objectives. For this analysis, we averaged the scores of the policymaking group into a single ranking of objectives, and used the mean values to design the future system.[11] The differences among individuals' scores and the differences between individuals' scores and the averaged ranking were preserved and considered in the sensitivity analysis. (See Chapter Six and Appendix D.)

---

[11]Keeney and Kirkwood (1975) demonstrate methods for assessing group preference. In essence, a group should be viewed as an entity. Two types of decisionmaking might take place. In one type, a single individual (e.g., the Assistant Secretary of Defense) or a small group makes the decision. This decisionmaker might incorporate the preferences of others into the decisionmaking process, but he or she ultimately must express a preference as an entity. In the second type of decisionmaking, an entire group is collectively responsible for the decision and has to decide, as an entity, on the weights. The weights could result from individual preference for each objective or, more likely, an arbitrated process that leads to a set of weights, with which each member agrees. According to Sackman (1974), this must be done to avoid manipulated group suggestion and achieve real consensus.

Our analysis follows the second type of decisionmaking, treating the preferences as a true group consensus. We believe this assumption is warranted because the working group, whose individual preference weights were similar to those of the policymaker group, reviewed its group mean and accepted it as a consensus. In our sensitivity analysis, we ignore the consensus to demonstrate the effects of individual weights.

Table 4.2 shows the ranking that resulted from the mean of the senior decisionmakers' preferences among the objectives.

Several important observations emerge from this ranking of the objectives. First, as the spacing in the table suggests, the results cluster into three main groups of objectives. "Keep costs reasonable" clearly stands separate from the rest of the objectives. The next four objectives—"provide career satisfaction," "emphasize cadre with military culture," "meet active experience needs," and "meet active skill needs"—emerge as a cluster of objectives with a similar level of im-

**Table 4.2**

**Prioritized Objectives**

| Ranking | Objective | Weight as Percent of 100 |
|---|---|---|
| 1 | Keep costs reasonable | 19 |
| | | |
| 2 | Provide career satisfaction | 13 |
| 3 | Emphasize cadre with military culture | 13 |
| 4 | Meet active experience needs | 12 |
| 5 | Meet active skill needs | 11 |
| | | |
| 6 | Inculcate culture prior to or at entry | 8 |
| 7 | Provide high opportunity to serve | 7 |
| 8 | Provide career opportunity | 6 |
| 9 | Meet reserve needs | 5 |
| 10 | Meet active grade needs | 3 |
| 11 | Be compatible with civilian careers | 2 |

NOTE: We used the application software *Expert Choice* to convert the numerical ratings from the interview questionnaire into the weighted rankings. *Expert Choice* provides a cardinal ranking for the complete list of objectives based upon the input of paired cardinal preferences obtained from the interview process. To do this, ratio weights among objectives at each level of the hierarchy are algebraically decomposed. In essence, among objectives being compared, the lowest in importance is given a value of 1 and others are given values as a factor of 1. For example, an objective deemed twice as important as the lowest one would get a value of 2 (and, if there were only two objectives, the ratio weights would be 1:3 and 2:3). The ratio weights derived are subsequently normalized to add to 1 at each level of their hierarchy (again, if there were only two objectives, 33 and 67 percent, which would be displayed in the table above as 33 and 67). The scores displayed in the table are the average of the ratio weights for that objective for the policy group interviews.

portance. The remaining six objectives are clearly less important than the top two groups (e.g., "keep costs reasonable" is nearly ten times more important than "be compatible with civilian careers").

A second observation is the relative tightness in the weighted preferences of these objectives. While "keep costs reasonable" is clearly at the head of the priorities, it does not command a majority of the total 100 percent.

Finally, it is interesting to note that all four of the broad considerations are represented in the top two clusters of objectives. "Keep costs reasonable" is included under "be consistent with public values." "Provide career satisfaction" is an objective under "foster careers." "Emphasize the cadre with military culture" is subsumed under "maintain a military culture." Finally, "meet active experience needs" and "meet active skill needs" are both part of "meet military manpower requirements."

Chapter Five

# LINKING ALTERNATIVES TO OBJECTIVES: EVALUATION MEASURES

In Chapter Three we recounted how we designed the alternatives for the OCMS. In Chapter Four we described how the group of policymakers, through a carefully constructed interview vehicle, rank-ordered the 11 objectives, or, to put it another way, we determined the intensity of the group's preference for each objective. The remaining task is to use the rank-ordered objectives to select alternatives that, in the aggregate, form the officer career management system.

This chapter explains the technique we used to evaluate alternatives against objectives and to take into account the varying policy effect of decisions. A generic explanation illustrates the methodology. Appendices D and E contain specific information about the objectives and alternatives of this study, and Appendix B gives an example of how we decided about one functional aspect, entry point. This chapter gives a general understanding of the methodology; those interested in the quantitative details of the methodology should turn to the appendices.

The chapter first describes how we normalized objectives so they could be compared and then how we accounted for the different weights policymakers assigned to them. It then describes how we accounted for the different policy effects of increments of change, and it concludes with a discussion of how we made judgments about how well an alternative accomplished a given objective.

## COMPARING ALTERNATIVES

The objective set contains very different things, from "keep costs reasonable" to "meet reserve needs" to "be compatible with civilian careers." We evaluate alternatives against very different objectives by determining how well the alternatives accomplish the objective set. We first determine how well they accomplish a given objective by using a normalized scale and, next, by how well they accomplish the set of objectives. We also determine the relative policy weight of an increment of change. Note that our methodology assumes independent objectives—i.e., the effect of alternative policies on one type (e.g., education) is independent of the alternative for another type (e.g., career length).

To normalize objectives, we develop a common scale for each. Our scale runs between 0 and 1, where 0 represents the complete failure to accomplish an objective and 1 full accomplishment. By plotting an alternative along this scale—that is, by deciding how well a given alternative accomplishes an objective—we can derive a quantitative value for it that enables us to compare it with other alternatives for accomplishing the same objective. Figure 5.1 depicts such a scale.

In Figure 5.1, we plot the degree to which alternative A accomplishes objective 1 of, hypothetically, five objectives. Scaling from the curve, we see that alternative A, the status quo, has a value of .5. Let's assume that we are considering two other alternatives, B and C, one of which accomplishes the objective better than the status quo and the other less well. We plot these on the same scale and derive a value for each, as shown in Figure 5.2.

This process results in scores for alternatives A, B, and C, respectively, of .5, .75, and .25. These scores tell us how well each alternative accomplishes objective 1. However, they tell us nothing about how well the alternatives achieve the other four objectives. The next step is to evaluate objectives A, B, and C against the other four objectives, deriving a score for each. Thus, in this example, we would evaluate three alternatives against five objectives and derive 15 values.

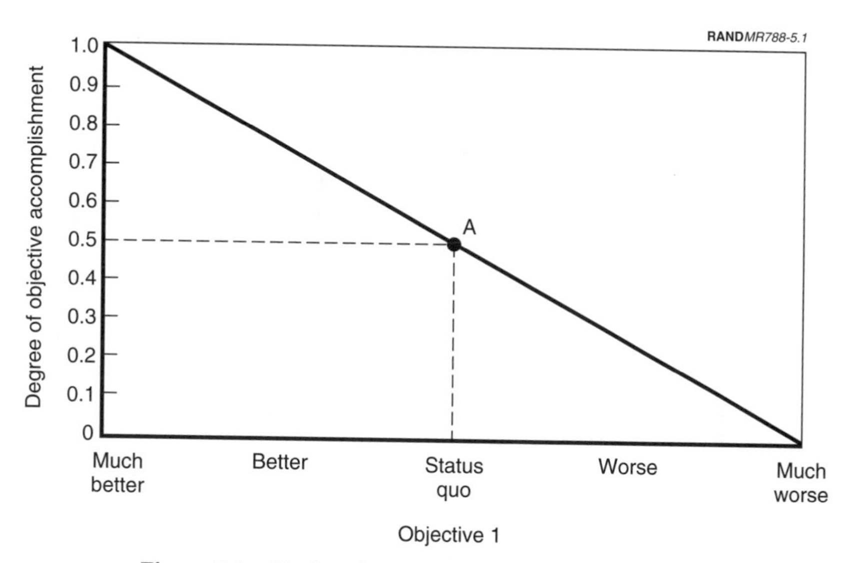


**Figure 5.1—Notional Example of a Normalized Scale**

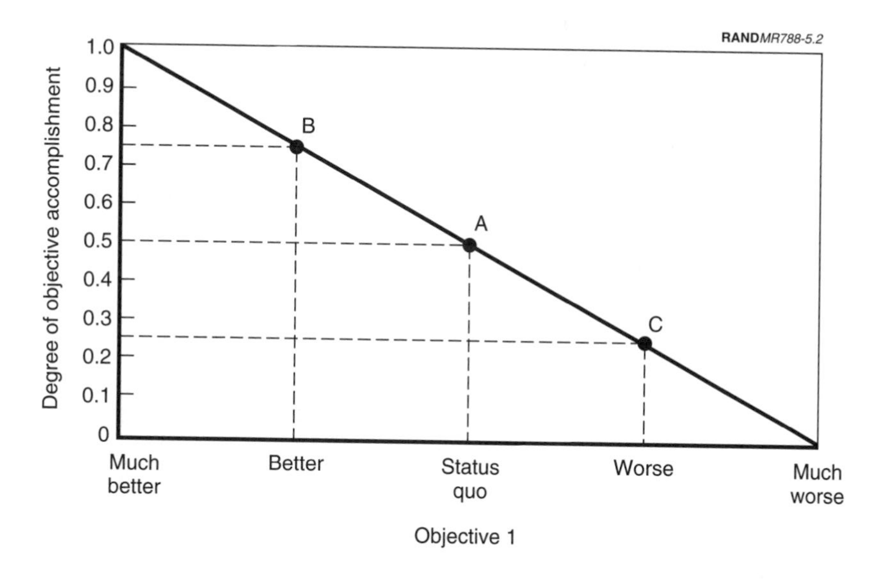


**Figure 5.2—Comparing Alternatives on a Normalized Scale**

Figure 5.3 shows how the three alternatives scored against the five objectives. We derive these values in a manner similar to the one we used for objective 1.

At this point, we could simply average values to determine which alternative best accomplishes the objective set—the one with the highest average score. But that approach assumes that all objectives have equal value, and we know from the interview with the policymakers that they regard some objectives as much more important than others. So we need to take the relative importance of the objectives into account before deciding which alternative best accomplishes all objectives. It may be that an alternative scores well at accomplishing a few important objectives but poorly on accomplishing many lightly weighted objectives. However, its high score on the objectives policymakers view as the most important may very well make it the alternative of choice.

We apply the relative importance of an objective by multiplying an alternative's score on that objective by the relative weight assigned to it by the policymakers. In this example, assume policymakers assigned objective 1 a relative weight of 19 percent. Figure 5.4 shows the result of applying the objective's weight to each alternative's score.

Figure 5.5 shows the result of applying all the policy weights to all the alternative scores. We compute scores for each alternative for each objective, apply the weight for each objective, and sum across all objectives. In this case, alternative A, with its score of .67, best meets all objectives, even though it did not best meet every objective.

RANDMR788-5.3

| | Objective | | | | |
|---|---|---|---|---|---|
| | #1 | #2 | #3 | #4 | #5 |
| Alternative A | .50 | .30 | .90 | 1.0 | .80 |
| Alternative B | .75 | .50 | .20 | .50 | .20 |
| Alternative C | .25 | .70 | .10 | .60 | .50 |

**Figure 5.3—Scoring Alternatives Against Objectives**

RANDMR788-5.4

| | Objective #1 |
|---|---|
| | Weight (%) = 19 |
| Alternative A | .5 x .19 = .09 |
| Alternative B | .75 x .19 = .14 |
| Alternative C | .25 x .19 =.05 |

Figure 5.4—Applying an Objective's Weight to Alternatives

RANDMR788-5.5

| | Objective #1 | Objective #2 | Objective #3 | Objective #4 | Objective #5 | |
|---|---|---|---|---|---|---|
| | Weight (%) | | | | | Total |
| | 19 | 25 | 30 | 11 | 15 | |
| Alternative A | .09 | .08 | .27 | .11 | .12 | .67 |
| Alternative B | .14 | .13 | .06 | .06 | .03 | .42 |
| Alternative C | .05 | .18 | .03 | .07 | .08 | .41 |

Figure 5.5—Determining the Best Alternative

## APPLYING POLICY WEIGHT—SINGLE-VALUE FUNCTIONS

The normalized curve depicted in Figure 5.1 implicitly assumes that policymakers remain indifferent to all increments of change, or, to state it differently, that every increment of change has an equal policy effect. But from our interviews and other research, we know that not to be the case. In general, policymakers tend to care more about a decision the further it moves away from the status quo. That is, they regard the policy implications as greater.

For example, assume that objective 1 in the example is "keep costs reasonable." Policymakers tend not to feel as strongly about small changes from the status quo as they do about large ones because small ones usually have smaller effects. An alternative that doubles costs would provoke a much stronger reaction than one that increases them by 5 percent.

We apply what we call "single-value functions" to modify the straight-line curve so that it better reflects the policy implications of decisions.[1] We do this by making judgments about the policy effect of moving away from the status quo. Typically, we consider five evaluation points for each objective:

- the status quo,
- better than the status quo,
- significantly better than the status quo,
- worse than the status quo,
- significantly worse than the status quo.

The specific definition of the evaluation point varies by objective. It has to be relevant both to the aspect and the objective. For example, in dealing with the aspect of entry point and the objective "keep costs reasonable," the dollar cost of acquiring an officer is relevant to both, and that becomes the basis for the evaluation. We determined end points by ensuring that they reasonably bounded the alternatives. In the case of acquiring an officer, the least expensive way of acquiring an officer became one end point because our previous study showed that it could not be done much more cheaply. A 20 percent increase over the status quo became the other end point because it was a substantial increase, and it was unlikely that policymakers would support costs much higher. Also, the specific values were less critical than their relation to each other. That is, the costs

[1]These functions are also referred to in the literature as "single dimensional value functions" or "single attribute value functions." We have adopted the abbreviated terminology in this text for simplicity's sake.

might not be precise, but as long as the direction of effect and magnitude were approximately correct, we could assess the effect on the objective accurately. We determined end points for each of the aspects and objectives in a similar way, that is, by defining a relevant evaluation point and then assuring ourselves that the range of the scale was broad enough to include a reasonable range of alternatives.

Our determinations of these single-value functions were informed by the data and analysis of the earlier report, the interviews, and the research for this study. We discuss how we used this information in more detail in the next subsection. However, we did not rely solely on our own assessment. We vetted the single-value functions with selected members of the policymaking group, with the entire working group, and with outside economics and decision analysis experts.[2] Their comments have been incorporated into the functions. We also conducted a sensitivity analysis by computing the values for alternatives both with and without the single-value functions. Few changed significantly. We thus believe the single-value functions provide a more accurate—if not significantly different—representation of the policy effect.

Figure 5.6 illustrates the effect of applying single-value functions to the normalized curve for objective 1, "keep costs reasonable."

The curve now reflects the policymakers' preference for alternatives that save money and aversion to those that increase costs. The status quo is no longer at the midpoint on the normalized scale; and the lower value (0.42) reflects the policymakers' dissatisfaction with maintaining status quo costs and their preference for cost savings. Correspondingly, alternative C receives a lower score because it represents a significant cost increase, and alternative B a higher one because it brings significant savings. Table 5.1 compares the scores with and without the single-value function. The difference between scores represents the policy significance of the change in costs from lower to higher.

---

[2] A department chairman from a university economics department and an expert in decisionmaking analysis on the faculty of the Air Force Academy.

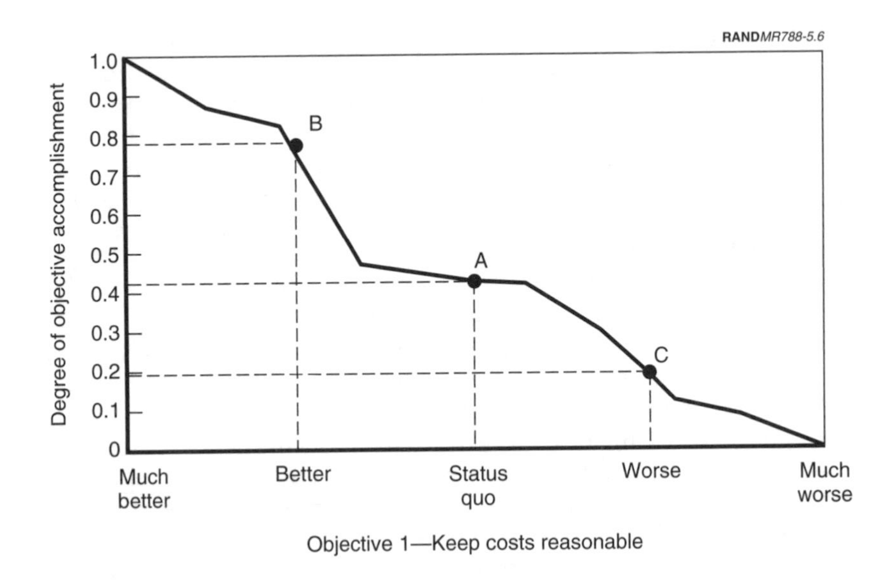


**Figure 5.6—Illustration of Single-Value Functions**

**Table 5.1**

**Effect of Single-Value Function**

| Alternative | Normalized Score | With Single-Value Function |
|---|---|---|
| A | .50 | .42 |
| B | .75 | .79 |
| C | .25 | .19 |

## DETERMINING HOW WELL AN OBJECTIVE IS ACCOMPLISHED

A question deferred from earlier in this chapter is how we determined the degree to which a given alternative accomplished an objective. This step is crucial to the scoring and, ultimately, to the OCMS that emerges. We could have done this—as other studies do—in the same way we determined the objective weights, by interviewing policymakers. However, we decided to base our evaluations on the data and analyses of the previous study. We had three types of data: quantitative, qualitative based on quantitative, and qualitative. However, we used each in a similar way. We ascertained the direction and the magnitude of the effect with respect to the status quo and converted them into qualitative numerical indicators relative to the status quo.

## Quantitative

We represented quantitative data directly. The quantitative data were represented by a curve, which of course contains an infinite number of points. For example, in evaluating the objective "provide career opportunity," the measure relative to the objective and aspect was "expected career length," which ranged from 11 to 19 years (see Table C.14). We depicted the data in six-month increments.

## Qualitative Based on Quantitative Data

When we had quantitative information on which we could base a qualitative judgment—i.e., better or worse than the status quo—we translated it into qualitative indicators, usually representing it by a ± 1 or 2 depending on the direction of the effect (i.e., better or worse), with the status quo as 0. As suggested above, we typically used five points. An example is evaluating the cost of acquiring an officer with respect to the objective of "keep costs reasonable." (See Table C.6.) Any alternative that cost more than the status quo would add expense and thus score lower on achieving the objective, and alternatives cheaper than the status quo would score higher. To illustrate, considerable data were collected in the earlier report regarding the cost of obtaining an officer from the different accession alternatives. Those data show, for instance, that it costs slightly over $21,000 (FY 1989 dollars annualized over an expected career length) to bring an officer into the service at the beginning of a career, what we call entry point 0. Other alternatives—lateral entry from reserves or prior service and lateral entry from civilian life—cost about $19,000 and $17,000, respectively. As mentioned above, the previous study showed $17,000 or $4000 less than the status quo probably represented the low end of costs for procuring an officer and thus became an end point. To ensure we captured a reasonable range of alternatives, we established the other end point by adding an equal amount to the status quo—$4000. Once the end points of the x-axis of a single-value function are designated, locating an alternative on the single-value function curve is straightforward.

However, exceptions to the five-point standard occurred. "Maintain compatibility with civilian careers" provides an example. We determined that compatibility was an absolute; that is, once an alternative was compatible, it could not be more compatible (there were degrees of incompatibility). So the scale contained only three points, as indicated in Table C.15.

## Qualitative

Qualitative research studies were also used as the basis of evaluative decisions. In these instances, the research indicated the directional effect particular changes would have on officer careers, with some sense of their magnitude. To make the most efficient and accurate use of these studies, we employed a discrete scale to evaluate alternatives based upon them. An example of this type of evaluation is vesting with respect to achieving the objective "be compatible with civilian careers." Our research into private sector pension plans shows that companies that have plans must, by law, vest their employees, either fully after five years or partially over a series of years, but with full vesting occurring by year seven. So we know that any no-vesting practice would be totally incompatible with civilian careers (that is, vesting occurs only at retirement eligibility) and that somewhere between five and seven years would be fully compatible. Thus, one end point became 20 years (the status quo in the military) and the other became four to six years. Knowing the end points, scaling alternatives between them is straightforward. Chapter Six describes the officer career management system that emerges from the evaluation of the alternatives.

Chapter Six

# DERIVING AN OFFICER CAREER MANAGEMENT SYSTEM

Knowing what decisions to make about each aspect of the four personnel areas allows us to derive an officer career management system. Reviewing the alternatives that score best in each of the personnel areas, we find that the line officer career management system that best satisfies all weighted objectives is characterized by high early turnover, by broader and deeper development, by a promotion system that is based on merit and tied to requirements, and by longer career lengths. Details of this system for each of the personnel functions appear below.

## ACCESSING

Certain entry practices best fit the preferences for objectives. For line officers, the alternatives that best support the objectives show that

- the career system should remain predominantly a closed system, that is, with entry at the beginning of a career;
- enlisted service acculturates well;
- a long payback period for pre-entry education and training should exist; and
- there should be a six-year period of initial tenure for those aspiring to longer careers.

## Entry at Year 0

Continuation of the existing practice of entering line officers at the beginning of a career emerges as best. However, lateral entry of those with either reserve service or prior active service also fits the stated objectives well. Both practices do well at meeting active component skill and experience needs and emphasizing those who possess military culture. Entry of experienced personnel saves some training expenses; a completely closed system provides greater individual career satisfaction.

## Enlisted Service Acculturates Well

Different methods of acculturating officers have different characteristics: educational or experiential, level of intensity, and duration. All the methods for acculturating before or immediately after becoming officers score well against preferred objectives. Acculturation based on experience of medium intensity and of long duration in the military as provided by enlisted service emerges as best across all objectives. However, that all methods do well suggests that they all have merit, and the services should continue to use some combination of them in their officer accession programs.

## Two-Year Payback for Each Year of Pre-Entry Education

Many officer entry programs provide a college education to those who aspire to be officers. These programs include the several service academies, ROTC, and various service programs that provide college degrees to enlisted aspirants. Generally, the cost of providing this education is amortized over a period of years through an initial service obligation that varies by source of commission and amount of expense involved. We evaluated payback periods of different lengths, and longer payback periods are increasingly preferred over shorter periods.[1] The objective "keep costs reasonable" is most significant in mandating longer payback periods. Longer paybacks

---

[1]We did not evaluate beyond a two-year period of providing education for a one-year payback period (2:1 ratio). Also not evaluated was the potential effect of a longer payback on supply.

amortize the investment costs best given less significant changes in the other objectives.

### Initial Tenure of Six Years

The obverse to required service is the amount of allowed service before involuntary separation of those who would prefer to stay for longer periods. Not everyone who wants to make the military a career will be able to do so. A desirable balance would be to have enough time to be certain that those who are allowed to continue in careers are those who meet service needs but not so much time that costs increase or individual satisfaction diminishes. A six-year tenure period (about two years beyond the current promotion point to O3) best satisfies all objectives.

## DEVELOPING

As officers move into career status and gain greater experience through assignments and more knowledge through education, the best practices for line officers are

- selection (augmentation) for career status after six years of service;
- some mandatory transfers to other skill groups for those who wish military careers but are not selected for line careers;
- longer assignments; and
- more military and civilian education.

### Selection for Career Status After Six Years

At what point should a service decide to continue line officers into career status? Sometime after the initial tenure period and by 10 years of service emerges as best in satisfying all objectives. Making the decision after the initial tenure period for officers but before 10 years of service best provides career satisfaction to individuals by removing uncertainty, yet separates officers from active service early enough to be useful in meeting reserve needs for junior officers.

### Some Mandatory Skill Transfers

Because it requires more junior officers to meet line officer grade and experience needs, some line officers who want to continue their line careers at higher grades will be unable to do so. However, directing the migration of certain numbers of such officers to careers in other skill groups such as support contributes significantly to emphasizing the military culture, to providing individual career satisfaction, and to meeting active skill needs.

### Longer Assignments by Up to Two-Thirds

Increasing average assignment lengths emerges as a better practice than either continuing current lengths or reducing them. Longer average assignments better achieve skill needs and reduce cost. However, making assignments too long can diminish career satisfaction, so there is a limit on how much assignment length should be increased.

### More Military and Civilian Education

We evaluated more education (up to two years more over a career), less education (up to two years less), and the current amount of education. While more education best fits preferences across all objectives, the differences are not large. More education costs more but contributes to meeting skill needs and to providing career satisfaction.

## PROMOTING

As line officers develop, they expect higher levels of responsibility. The mechanisms that accomplish this best against the set of objectives are

- a promotion zone based on time in grade (TIG) given minimum time in service (TIS);
- a "long" zone of selection that varies from three to eight years;
- selective opportunity based on actual requirements; and
- selective continuation.

## Promotion Based on TIG Given Minimum TIS

This alternative, which allows officers to be promoted more quickly than a time-in-service alternative, ranks first against objectives because it provides the greatest individual career satisfaction and is more compatible with civilian careers, which generally do not tie promotion qualification to amount of service. Time in service also ranks highly against objectives because it contributes to meeting experience needs. (That is, it helps to maintain a grade and length of service tie.) The TIG alternative, which allows officers to be promoted most quickly, ranks last because it does not satisfy well the objective for active military experience.

## Wide Zone with Selective Opportunity and Selective Continuation

As discussed earlier, we treat several dimensions of the promotion system together, which sets up 10 alternatives to be evaluated. The current DOPMA system has a narrow one-year zone for selecting most officers to be advanced, fixed opportunity for promotion (although there are minor service and annual variations, the general policy is to promote 50 percent of O5 to O6; 70 percent of O4 to O5; and 80 percent of O3 to O4), and selective continuation (certain officers may be kept in service even if not promoted). Such an alternative does not do well for the future against preferred objectives. The alternative that fares best allows officers to be considered for promotion selection over multiple-year periods (three years for O3 to O4, five years for O4 to O5, and eight years for O5 to O6), is requirements based in that it advances officers only when vacancies exist (and thus promotion opportunity will vary from year to year with numbers of officers eligible for promotion and as requirements change), and allows officers who are not selected for promotion to be selectively continued based on each service's needs. This alternative is best against objectives dealing with meeting skill and grade needs and "keep costs reasonable." In general, all alternatives that use selective promotion opportunity and either a broad or completely open zone fare best; a middle group of alternatives uses the broad zone with fixed promotion opportunity; and the last ranked group of alternatives uses the narrow one-year zone.

## TRANSITIONING

For line officers, the personnel management practices that emerge as best satisfying preferred objectives are

- vesting;
- no intermediate tenure, no voluntary separation incentives;
- high turnover early in service;
- modified up-or-out;
- longer careers; and
- later retirement annuities.

### Vesting Between Four and Nine Years of Service

Our assessment of the various vesting alternatives is informed by the Asch-Warner studies, and our results are consistent with their more detailed assessment of this issue.[2] In particular, they have concluded that vesting with an old-age annuity[3] early is consistent with private-sector practice and may create fairness but does not come cheaply. However, as they state, the practice might induce useful separations or lead to a grade structure with fewer field-grade officers. In our

---

[2]For example, see Beth J. Asch and John T. Warner, "Should the Military Retirement System Be Reformed?" in J. Eric Fredland, Curtis L. Gilroy, Roger D. Little, and W. S. Sellman (eds.), *Professionals on the Front Line: Two Decades of the All-Volunteer Force*, Washington, D.C., Brassey's, 1996. See also Beth J. Asch and John T. Warner, *A Policy Analysis of Alternative Military Retirement Systems*, RAND, MR-465-OSD, 1994.

[3]Vesting, the right to share in some future fund after certain periods of employment, can be implemented in several ways. The Employee Retirement Income Security Act, which regulates such plans, does not apply to plans established or maintained by the U.S. government. We considered early vesting without an immediate annuity in our study at early points (4–9 years) and later points (10–14) because it may be useful in meeting certain objectives. We make no economic assertions that our point of vesting is theoretically correct; we assert only that it may be useful to achieve certain objectives along the lines that Asch and Warner suggest (e.g., it is equitable with private-sector practice). For instance, in unpublished research, Asch and Warner conclude that a package of vesting and separation incentives is cost-effective. We have not considered the full range of options evaluated by Asch and Warner. Some may be better than the alternatives we consider here. Our conclusions are consistent with their analysis: as weight is increased on cost relative to other objectives, not vesting emerges as preferred.

assessment, vesting between four and nine years of service is more costly but does contribute to satisfying objectives such as "meet reserve needs," "meet active skill needs" (reduce imbalances), and "be compatible with civilian careers." Because of weights placed on objectives, such vesting ranks slightly higher than not vesting at all.

## No Intermediate Tenure; No Voluntary Separation Initiatives

Our assessment is that, given the weight on objectives, neither tenure (a promise for continued service given promotion to certain grades) nor incorporating separation incentives as a matter of due course is a useful practice around which to design a career management system. We evaluated what we called "transitional ability"—the ability of the system to react to change—and no constraints and no incentives were ranked highest.

## Modified Up or Out

We evaluated two alternatives for actions to be taken when an officer fails promotion selection: separation (as in the current system) and continuation in skill or grade. The alternative that ranks highest against all objectives is that of continuation in skill and grade. Separation ranks lowest. However, from the earlier promotion evaluation, the type of continuation that is best is selective rather than fixed. That is, rather than arbitrarily fixing the opportunity for continuation (at 100 percent or some lesser percentage), the number of officers to be continued would be up to each service and subject to vacancies in skill and grade. This type of vacancy system mandates decisions about individual officers rather than adherence to rules about categories of officers. Thus, we describe it as a modified up-or-out system. The amount and timing of the separation depend on service decisions about particular individuals.[4]

---

[4]As we modeled outcomes, we allowed some O4s to serve to about 25 years, but most had left after not being selected for O5 in the long zone system. We allowed O5s and O6s to continue until the maximum allowed career length. This is only one potential design for a line officer system; a service could choose to implement it differently.

## High Turnover Early in Career; Longer Careers; Later Retirement Annuities

As explained earlier, because turnover (separation rates in first ten years), maximum career length, and retirement annuity point are related, we evaluated them jointly. The choices for each led to some 54 alternatives to be evaluated. After evaluation, the alternatives grouped in ways that enabled us to draw overall conclusions about them. The conclusions are starkest at the extremes. Those alternatives that rank most highly are characterized by high turnover in the first ten years of service with longer allowed career lengths (up to 40 years) and with later retirement annuities (immediate retirement annuities delayed to within five years or less of the allowed career length). The high-turnover-early systems always rank highest except when they are combined with the shortest career lengths or earliest retirements. As a general rule, any alternative that allows immediate retirement annuities after only 15 years of service ranks lowest. Most alternatives that allow retirement annuities at 20 years of service also ranked low, the exceptions being when 20-year annuities are tied to longer careers (35+ years) and high turnover early in the career.

# COHERENCE: THE SHAPE OF A FUTURE CAREER

Our method for determining preferences for objectives was based on achieving external fit with the likely future environment. A career management system should also hew to the important principle of coherence or internal fit. The specific practices chosen for each aspect should complement one another and support (and be supported by) other human resource practices such as compensation. The career management system should have an underlying logic, and the various practices should fit together. If the practices do not fit, one or another must change.[5] The system that emerges from our

---

[5]To resolve any issues of coherence, we asked in our interviews which of the personnel management functions was most important in designing a career management system. The decisionmakers as a group thought that developing was the most important personnel function and transitioning was the least important. Had the career management system emerging from the prioritized objectives appeared to suffer from internal coherence problems, we could have based resolving decisions upon the score of the alternative and the importance of the function it supported. For example, an alternative that develops officers well would likely remain, whereas an alternative that

objectives-based design meets the coherence test. Figure 6.1 depicts the shape of the objectives-based career for a typical line officer and compares it with DOPMA. We modeled the promotion and flow characteristics for each service and skill group to assess their feasibility.[6] This new career "shape" resembles DOPMA for some services. For example, high turnover early has generally characterized Marine Corps officer management.

However, other aspects differ. (This profile is for one particular service; the shape is similar for the other services with minor differences because of different numbers in different skill groups and retention patterns.) Because the new career allows for more cross flow from the line to other skill groups such as specialist or support,

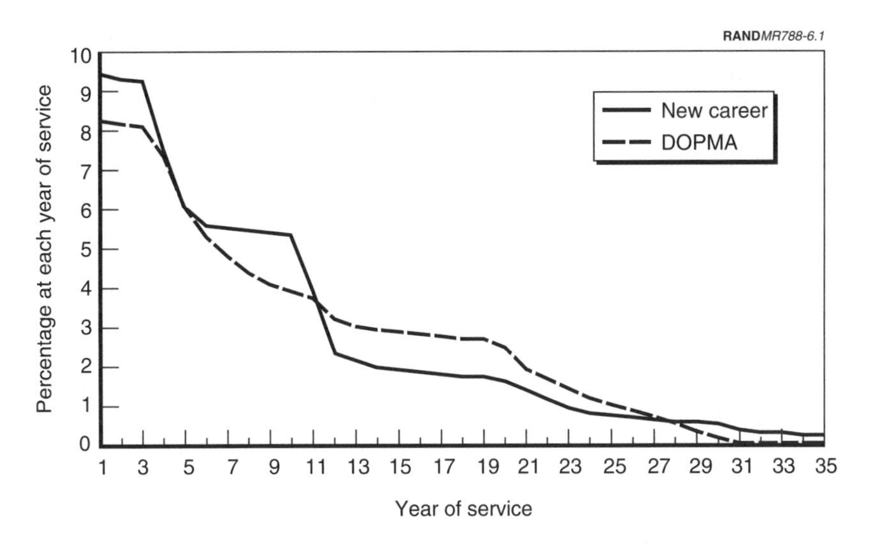


**Figure 6.1—Shape of a Future Career for Line Officers**

transitions officers only slightly better than the next alternative might be sacrificed for the sake of a coherent system, given the relative importance of developing and transitioning officers.

[6]See Appendix E for a more detailed explanation.

proportionally more officer accessions enter the line than is now the case.[7]

College graduates or those with a potential to complete a college degree will enter as officers from many different sources, as they do now. They will be imprinted with military culture—values, attitudes, and beliefs—in various ways: through education in service academies and in ROTC programs and, increasingly, through experience in the enlisted force. Transformation from civilian life will receive considerable emphasis. Because the enlisted force will be more educated and better motivated than ever, proportionally more of the officer accessions will be noncommissioned officers (NCOs) and petty officers.

Entering a military smaller by one third than it was during the cold war, junior officers will find competition rigorous for the best assignments. The services will want reasonably long periods of required service (somewhat longer than now) to amortize investments in education and training but will provide reasonably long periods of initial service (somewhat longer than now) to evaluate those with the best prospects for full careers.

By the sixth year of service, proportionally more initial entrants than now would have completed service and separated or would have migrated to another skill group for continuing service. By the tenth year of service, final decisions would have been made about who enters into a potentially long service career, and fewer people enter such careers. Proportionally more officers than now will separate, which gives the system its high-turnover-early shape. These separations will be voluntary as payback periods for college education or advanced initial training expire and involuntary as a result of not being selected for career status. Other separations from the line skill group will occur, but these officers might be directed to careers in other skill groups. Both voluntary and involuntary separations will be sup-

---

[7]The direct accessions into other skill groups are not shown in this new career line officer profile. The overall proportion of accessions for all skill groups is about the same as it is for DOPMA. The proportion that enter the line versus those that enter other skill groups would vary by service because service requirements for officers at lower grades in each skill group are different.

ported by a vesting system that promises future payment based on past service and eases the impact of involuntary separation.

Although proportionally fewer officers than now will continue into career status, once in those careers, they can look forward to greater stability in assignments, to additional assignments in their career path, and to more military and civilian education. Moreover, higher proportions of them than now will be able to serve in important positions, with a corresponding effect on compensation, status, and promotions. The promotion system will advance the most meritorious somewhat more quickly and will provide longer periods over which to judge merit for promotion. In a significant departure from current practice, fixed percentages of annual promotion opportunity will not be promised in advance; officers will be promoted as vacancies arise, which means opportunity will vary as requirements change or as the number of officers eligible for promotion changes. However, depending on the service (because of requirements and retention differences), promotion prospects will remain good because fewer officers will enter into careers and a modified up-or-out system will continue. In general, some few O4s might continue to a later retirement, but most will be separated if not promoted. Generally, O5s and O6s will continue to the retirement annuity point, which will be later than the current 20 years. Less-pronounced separation occurs at 20 years of service because of vesting and the elimination of the 20-year immediate annuity. This will fit with the allowed career length that will extend to at least 35 years.

## FLEXIBILITY ACROSS SERVICES

Mainly because the career described above is based on requirements for officers (a "vacancy"-based system), but also because services retain people at varying rates, differences are likely to occur among services. In our earlier study, we outlined several forms that requirements for officers could take in the future, and our methodology can accommodate requirement changes. However, the objectives were weighted based on the military operating environment—missions, organizations, and technology—and personnel requirements evolving in a straightforward way. Even so, differences will occur among services because of grade, skill, and continuation differences without making any of the more significant changes to numbers, grades, and skills that could be made.[8] For example, the Marine Corps has the highest line content and the fewest professionals. The Air Force and Navy have double the specialists of either the Army or Marine Corps. Specialist and professional skills have the most O4–O6 requirements. Line officers have the fewest O4–O6 requirements. Within the line skill group, the Navy has the most O6 and the Air Force the most O5 requirements. The Army and Marine Corps have proportionately similar line grade structures.

Would such differences be acceptable? That is, could there be a move away from an officer management system that strives for uniformity toward one that allows for more flexibility across services? Part of our interviews with the policymaking group dealt with that question. Figure 6.2 shows the result. The arrows indicate the policymaker consensus on a flexibility-uniformity spectrum.

In general, decisionmakers preferred the most uniformity across services in the transitioning function. When and under what condi-

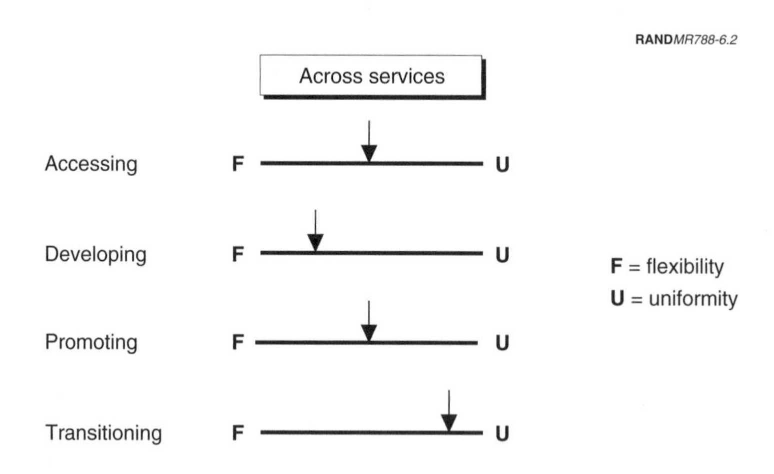


**Figure 6.2—Policymaker Preference for Flexibility or Uniformity Across Services**

---

[8]One requirement option significantly reduced the numbers of O4–O6, which is fitting with organizational changes and a reduction in mid-level staffing in the private sector.

tions line officers can separate should be most consistent across services. Developing—how line officers are assigned and experienced—is the function to which the most flexibility should apply. Accessing practices did not evoke strong preferences between flexibility and uniformity and promoting shaded slightly toward uniformity.

## SENSITIVITY OF DESIGN TO WEIGHTS ON OBJECTIVES

To determine if the career management system would change if the weights given objectives change, we conducted a sensitivity analysis by varying the objective weights from 0 to 100.[9] Some of the alternatives emerge as robust across the spectrum of weights. For these, we can conclude that data and analysis determine the best alternative; such practices should figure into any officer career management system. But some alternatives are sensitive to changes in weights on objectives. For these, the choice depends on priorities for objectives and require policymaker decisions.

During our sensitivity analysis, we gained an important insight about the effects of changed weights. The choice of an alternative is sensitive not only to the increased weight placed on one particular objective but is also sensitive to which other objectives lose weight. The set of weights of all objectives matters more than the weight of one.

To illustrate the effect of different objective weights, we selected examples of members of the policymaking group whose choices differed substantially from the consensus. We then applied the methodology using that policymaker's weights. Of the 11 objectives, seven were selected as a top priority by at least one member of the policymaker group, and these seven sets of objective weights were used in this sensitivity analysis. Table 6.1 summarizes the effect of the different objective weights. (See also Appendix D.)

---

[9]The detailed sensitivity analysis in Appendix D includes determination of break points—when conclusions change because of mathematical change in objective weights. Here we are more concerned with relevant policy choices by decisionmakers and portray sensitivity to individual decisionmaker weights on objectives rather than the mean senior group weights used for the main analysis.

**Table 6.1**

**Effect of Changing Objective Weights**

| When highest weight is given to | New objective weight (%) | It causes changes in | | | |
|---|---|---|---|---|---|
| | | Accessing | Developing | Promoting | Transitioning |
| Keep costs reasonable | 59 | • System becomes open<br>• More emphasis on accession programs with no acculturation | • Less education | • Based on TIS | • No vesting |
| Provide career satisfaction | 49 | • Shorter payback ~1:1 | | • Always eligible | • Intermediate tenure |
| Emphasize cadre with military culture | 31 | • Shorter payback 0.5:1 | | • Always eligible | • Intermediate tenure |
| Meet active experience needs | 29 | | • Longer assignments | • Based on TIS<br>• Always eligible | • No vesting<br>• Separation incentive<br>• Moderate turnover early |
| Meet active skill needs | 40 | | | • Based on TIS<br>• Always eligible | • No vesting |
| Inculcate culture prior to or at entry | 48 | • More emphasis on high intensity, long duration | • Less education | | |
| Provide high opportunity to serve | 20 | • Shorter payback 1:1<br>• 4 year intermediate tenure | | | • Separation incentive |

The officer career management system developed in this analysis depends on the set of prioritized objectives. Thus, the resulting system and the objectives themselves should be considered together in the process of implementation. To the extent that future policymakers agree with the weights placed upon different objectives, the career system that matches these preferences will continue to be appropriate. Changing priorities on objectives will change the system, but the new system can be easily determined with the methodology and framework provided.

The prioritized objectives might change for three reasons. First, a different set of senior policymakers could have considerably different priorities. Second, a change in the assumed future environment could produce different preferences for desired ends. This study assumed straightforward evolution to the year 2005. Dramatic domestic or international developments could have a considerable impact on the future of the U.S. military and the appropriate career management system. Thus, external fit could be jeopardized by significant scenario changes from the one we analyzed. Finally, policymakers could change their perspective about the evaluation measures for different objectives. For example, when we asked the policymakers for the relative importance of "keep costs reasonable," they gave a wide range of answers. This might have resulted from real differences or from unique perspectives on cost differences. If a policymaker envisioned cost variation on a scale of millions, he might have placed less weight upon the importance of this cost than a policymaker who routinely deals with large cost figures and was imagining significant cost savings or increases on a scale of billions.

Regardless of whether the objective weights change and the resulting career management system evolves, this study provides a means to articulate the dependent relationship between what a career system is trying to accomplish (the ends) and the practices to gain them (the means). In addition, the sensitivity analysis should provide an excellent preview of how changes in objectives will contribute to subtly or considerably different career management systems. Of special interest in any case should be the set of career management system practices that emerge consistently, regardless of the weight placed upon the objectives.

Chapter Seven

# CONCLUDING OBSERVATIONS

This study has defined and prioritized the objectives of current policymakers for future officer career management. The objectives that emerge from this process as most important for the likely future environment are "keep costs reasonable," "provide career satisfaction," "emphasize cadre with military culture," "meet active experience needs," and "meet active and skill needs."

Multiobjective decision analysis enabled us to identify promising alternatives to the current career officer management system. The new system satisfies the policymakers' objectives with high early turnover, broader and deeper individual development, a promotion system better based on merit and personnel requirements, and longer careers. It depends heavily upon the priorities accorded the objectives and on our application of current knowledge about the effects of career management on alternatives. Should these priorities change significantly or new evidence emerge on the effects of alternatives, the system might no longer be appropriate.

The value of this study goes beyond the specification of a single officer career management system. It provides insights into and a framework for addressing change, and directs future discussion to the motivations for change and the priorities of the decisionmakers contemplating it. In addition, the sensitivity analysis and the career systems proposed for each of the four skill groups provide an understanding of how the proposed system should change if the supporting priorities change. Future work might further assess tradeoffs (objective weights), develop better estimates of the effects of alternatives on objectives, and investigate how the system might respond in scenarios other than the one of straightforward evolution we use. Further work might also investigate risks associated with potential OCMSs.

Appendix A

# INTERVIEW VEHICLE

This appendix contains the interview vehicle used to obtain the policymakers' preferences for the objectives, their views on the relative importance of each of the four career management functions (accessing, developing, promoting, and transitioning), and their preferences for uniformity or flexibility across the services and across the skills in the career management system. Their responses to the questions were input and calculated using the software *Expert Choice.*

The interviews were conducted individually and followed the interview vehicle precisely when asking the questions. Any questions during the interview were resolved before returning to the interview vehicle.

## OFFICER CAREER MANAGEMENT SYSTEM GOALS AND OBJECTIVES

You have been asked by Mr. Pang to participate as a member of a [working group] [senior group] of decisionmakers during the process of determining the next officer management system. As part of our research, we have been asked to highlight ideas and concepts about career management. We are proceeding from the conclusions of our previous study that the first step in this process is to specify the ends that the career management system needs to accomplish. To assist in deciding about the next officer career management system, we are conducting structured interviews to determine how decisionmakers value and rank in importance certain objectives related to **future** officer career management. For us the future is about the year 2005, and the world has evolved in straightforward ways from the present.

We believe the ultimate goal of the career management system is to **develop and deploy the nation's human capital against national security needs**. What the career management system strives to do is to link human performance and satisfaction to the goals and challenges of the military organization. We are most interested in a subset of the nation's human capital—those who have the ability and motivation to become officers. In earlier discussions, we established that the military officer is expected to be of high quality, by which we mean someone who is intellectually and physically vigorous. Officers are also expected to be conscientious in that they willingly expend effort toward goals, and versatile in that they can accomplish multiple tasks now and have the capability to learn—can adapt to the future—as the environment changes. As you answer these questions, keep in mind it is this "high quality" officer we are discussing.

Certain objectives must be accomplished to reach the overall purpose. We will be asking you about **four broad considerations**: **(1)** meeting manpower requirements; **(2)** being consistent with the public's values about military service; **(3)** maintaining a military culture; and **(4)** fostering military careers.

This interview includes 33 questions, and we have found that it takes about 45 minutes. Your responses will be tabulated and grouped with the others being interviewed. After we have had time to process the responses, we will meet with the group to discuss further the responses and future officer management issues.

I will ask you to make a series of pair-wise comparisons that ask your preferences among these several objectives. We are interested in **your personal beliefs** and underlying philosophy, rather than an institutional view, regarding these issues. I will clarify what I mean by the terms in each question, but please ask me to elaborate further as needed. Also, besides your preference for an objective I will ask you how strongly you feel about it. You have a scale before you that can help you determine preference. If you believe both are equally important, that is a valid choice as well.

Any questions before we begin?

The first broad consideration for officer career management is to be consistent with the public's values about military service. Our belief is that the public cares strongly about two issues: keeping costs reasonable and providing a high opportunity to serve the nation in the military as officers and maintaining an officer force reasonably representative of society.

1. **Do you consider keeping costs reasonable more or less important to the public than having a high opportunity for individuals to serve as officers?**

1=EQUAL 3=MODERATE 5=STRONG 7=VERY STRONG 9=EXTREME

| 1. | REASCOST | 9 8 7 6 5 4 3 2 1 2 3 4 5 6 7 8 9 | OPPSERVE |
|---|---|---|---|

The second broad consideration for officer career management is to maintain a military culture. We have been told by many that this is important. Culture includes the values, attitudes, and beliefs that are held by people in the organization. In this case we are asking you to rate in importance certain objectives dealing with how best to emphasize and preserve the military culture.

2. **Which do you consider more important for maintaining military culture: placing emphasis on providing exposure to it and inculcating its values prior to entry as officers or placing emphasis on the cadre of experienced officers who possess and pass along the culture**?

1=EQUAL 3=MODERATE 5=STRONG 7=VERY STRONG 9=EXTREME

| 2. | PREENTRY | 9 8 7 6 5 4 3 2 1 2 3 4 5 6 7 8 9 | INCAREER |
|---|---|---|---|

The third broad consideration for officer career management is to foster careers. This objective has imbedded a big assumption—careers as military officers are desirable. Given that, there are three objectives that matter: career satisfaction, career opportunity, and reasonable compatibility with private sector careers. Career satisfaction comes about from such things as having work that is challenging and responsible, from having adequate compensation to include promotion opportunity, and from being able to accommodate family responsibilities. Career opportunity is being able to serve in a career for a reasonably long period of time. Lastly, compatibility with civilian organization career management practice in the United States is important because the military is part of society and should not be completely divorced from broad societal trends and concerns about careers. Thus, military officer careers should have some similarity with their civilian counterparts. I will now ask which of these—career satisfaction, career opportunity, or civilian career compatibility—you believe is more important for fostering careers.

3. **For fostering careers, do you consider providing career satisfaction or providing career opportunity more important?**
4. **Do you consider providing career satisfaction more or less important than maintaining compatibility with civilian careers?**
5. **Do you consider providing career opportunity more or less important than maintaining compatibility with civilian careers?**

1=EQUAL 3=MODERATE 5=STRONG 7=VERY STRONG 9=EXTREME

| | | | | | | | | | | | | | | | | | | | |
|---|---|---|---|---|---|---|---|---|---|---|---|---|---|---|---|---|---|---|---|
| 3. | SATSFCTN | 9 | 8 | 7 | 6 | 5 | 4 | 3 | 2 | 1 | 2 | 3 | 4 | 5 | 6 | 7 | 8 | 9 | OPRTUNTY |
| 4. | SATSFCTN | 9 | 8 | 7 | 6 | 5 | 4 | 3 | 2 | 1 | 2 | 3 | 4 | 5 | 6 | 7 | 8 | 9 | COMPATBL |
| 5. | OPRTUNTY | 9 | 8 | 7 | 6 | 5 | 4 | 3 | 2 | 1 | 2 | 3 | 4 | 5 | 6 | 7 | 8 | 9 | COMPATBL |

The fourth broad consideration deals with meeting manpower requirements in support of national security. This is usually stated as having the right grade mix, skill mix, and experience mix to meet needs of active users and providing reasonably junior prior active service officers to meet reserve component needs. (Typically this is between three and seven years of service.) I would like to get your preference between meeting needs of active users and meeting reserve component needs. Which do you believe is more important for meeting overall military manpower requirements?

6. **Do you consider meeting grade, skill, and experience needs of active component users more or less important than meeting needs of reserve component users for reasonably junior prior active service officers?**

1=EQUAL 3=MODERATE 5=STRONG 7=VERY STRONG 9=EXTREME

| | | | | | | | | | | | | | | | | | | |
|---|---|---|---|---|---|---|---|---|---|---|---|---|---|---|---|---|---|---|
| 6. | MEETACTV | 9 | 8 | 7 | 6 | 5 | 4 | 3 | 2 | 1 | 2 | 3 | 4 | 5 | 6 | 7 | 8 | 9 | MEETRESV |

On the active component side I want to know whether you believe it is more important to meet grade, skill (e.g., occupation), or experience needs. What contributes most in your eyes to the success of organizations in the national security establishment who use officers—getting the grade mix right, the skill mix right, or the experience mix right?

7. **Do you consider getting the right experience mix more or less important than getting the right skill mix?**
8. **Do you consider getting the right experience mix more or less important than getting the right grade mix?**
9. **Do you consider getting the right skill mix more or less important than getting the right grade mix?**

1=EQUAL 3=MODERATE 5=STRONG 7=VERY STRONG 9=EXTREME

| | | | | | | | | | | | | | | | | | | | |
|---|---|---|---|---|---|---|---|---|---|---|---|---|---|---|---|---|---|---|---|
| 7. | EXPRMIX | 9 | 8 | 7 | 6 | 5 | 4 | 3 | 2 | 1 | 2 | 3 | 4 | 5 | 6 | 7 | 8 | 9 | SKILMIX |
| 8. | EXPRMIX | 9 | 8 | 7 | 6 | 5 | 4 | 3 | 2 | 1 | 2 | 3 | 4 | 5 | 6 | 7 | 8 | 9 | GRADMIX |
| 9. | SKILMIX | 9 | 8 | 7 | 6 | 5 | 4 | 3 | 2 | 1 | 2 | 3 | 4 | 5 | 6 | 7 | 8 | 9 | GRADMIX |

Next, I want to ask you which of the broad considerations—meeting manpower requirements, being consistent with the public's values about military service, preserving a military culture, or fostering careers—is most important to meet the overall goal of developing and deploying the nation's human capital against national security needs.

10. **Do you consider meeting manpower requirements more or less important than being consistent with the public's values about military service?**
11. **Do you consider meeting manpower requirements more or less important than maintaining a military culture?**
12. **Do you consider meeting manpower requirements more or less important than fostering careers?**

**13. Do you consider being consistent with the public's values about military service more or less important than maintaining a military culture?**

**14. Do you consider being consistent with the public's values about military service more or less important than fostering careers?**

**15. Do you consider maintaining a military culture more or less important than fostering careers?**

1=EQUAL 3=MODERATE 5=STRONG 7=VERY STRONG 9=EXTREME

| | | | |
|---|---|---|---|
| 10. | MPRREQ | 9 8 7 6 5 4 3 2 1 2 3 4 5 6 7 8 9 | PUBVAL |
| 11. | MPRREQ | 9 8 7 6 5 4 3 2 1 2 3 4 5 6 7 8 9 | MILCULT |
| 12. | MPRREQ | 9 8 7 6 5 4 3 2 1 2 3 4 5 6 7 8 9 | FOSTCAR |
| 13. | PUBVAL | 9 8 7 6 5 4 3 2 1 2 3 4 5 6 7 8 9 | MILCULT |
| 14. | PUBVAL | 9 8 7 6 5 4 3 2 1 2 3 4 5 6 7 8 9 | FOSTCAR |
| 15. | MILCULT | 9 8 7 6 5 4 3 2 1 2 3 4 5 6 7 8 9 | FOSTCAR |

## PERSONNEL FUNCTIONS

This portion begins by querying you about the relative importance of four personnel functions that make up a personnel system: accessing, developing, promoting, and transitioning.

Accessing includes policies pertaining to such matters as initial active duty commitments; use of different sources for officers (e.g., academies, ROTC, OCS/OTS, etc.); the point in a career profile at which new entrants might enter into careers; and the standards (e.g., education, physical, aptitude, moral) officers must meet at entry. [Who; How; When]

Developing policies include the length and number of assignments in an officer's career; the nature of those assignments, usually represented by a career path; and the education (both military and civilian) needed.

Promoting includes policies pertaining to promotion opportunity, (probability of promotion to each grade); promotion timing (expected time in service before promotion to particular grades or the time in each grade between promotions); and the basis for promotion (e.g., emphasis merit or seniority).

Transitioning includes policies about tenure (limits on involuntary separation that would protect the individual); minimum service required before payment of an immediate retirement annuity; the amount of service required before an officer is vested for some future retirement annuity; and the maximum service allowed prior to mandatory retirement. [Who; How; When]

**When considering the overall goal of developing and deploying the nation's human capital against national security needs,**

1. **Do you believe that accessing is more or less important than developing?**
2. **Do you believe that accessing is more or less important than promoting?**
3. **Do you consider accessing is more or less important than transitioning?**

**1=EQUAL 3=MODERATE 5=STRONG 7=VERY STRONG 9=EXTREME**

| | | | | | | | | | | | | | | | | | | |
|---|---|---|---|---|---|---|---|---|---|---|---|---|---|---|---|---|---|---|
| **1.** | **ACCESSING** | 9 | 8 | 7 | 6 | 5 | 4 | 3 | 2 | 1 | 2 | 3 | 4 | 5 | 6 | 7 | 8 | 9 | **DEVELOPING** |
| 2. | **ACCESSING** | 9 | 8 | 7 | 6 | 5 | 4 | 3 | 2 | 1 | 2 | 3 | 4 | 5 | 6 | 7 | 8 | 9 | **PROMOTING** |
| 3. | **ACCESSING** | 9 | 8 | 7 | 6 | 5 | 4 | 3 | 2 | 1 | 2 | 3 | 4 | 5 | 6 | 7 | 8 | 9 | **TRANSIT'NG** |

4. **Do you consider developing more important than promoting?**
5. **Do you consider developing more important than transitioning?**
6. **Do you consider promoting more important than transitioning?**

**1=EQUAL 3=MODERATE 5=STRONG 7=VERY STRONG 9=EXTREME**

| | | | | | | | | | | | | | | | | | | | |
|---|---|---|---|---|---|---|---|---|---|---|---|---|---|---|---|---|---|---|---|
| **4.** | **DEVELOPING** | 9 | 8 | 7 | 6 | 5 | 4 | 3 | 2 | 1 | 2 | 3 | 4 | 5 | 6 | 7 | 8 | 9 | **PROMOTING** |
| **5.** | **DEVELOPING** | 9 | 8 | 7 | 6 | 5 | 4 | 3 | 2 | 1 | 2 | 3 | 4 | 5 | 6 | 7 | 8 | 9 | **TRANSIT'NG** |
| **6.** | **PROMOTING** | 9 | 8 | 7 | 6 | 5 | 4 | 3 | 2 | 1 | 2 | 3 | 4 | 5 | 6 | 7 | 8 | 9 | **TRANSIT'NG** |

## MANAGEMENT PRINCIPLES

This portion is intended to explore your values and preferences regarding specific management principles that will influence the design of an officer career management system. I will ask you to state your preferences about uniformity and flexibility in future officer management systems.

In particular, these questions address the desirability for uniformity or flexibility across services and across individual skills. Uniformity by service means that all services would adopt common policies; flexibility across services means that services could adopt unique policies. Uniformity by skill means that all skill groups would have common policies. Flexibility by skill means that different skill groups could have different policies. Please consider "skill groups" to refer to the four career types we outlined in our previous study: line (unique military skills), specialist (line skills that require additional education or technical expertise), support (skills with a civilian counterpart), and professional (e.g., medical or legal).

7. **Do you prefer uniform accession or flexible accession policies across services?**
8. **Do you prefer uniform or flexible accession policies across skills?**
9. **Do you feel more strongly about this issue across services or across skills?**

1=EQUAL 3=MODERATE 5=STRONG 7=VERY STRONG 9=EXTREME

| ACCESSION | | | |
|---|---|---|---|
| 7. | UNIFSVC | 9 8 7 6 5 4 3 2 1 2 3 4 5 6 7 8 9 | LEXSVC |
| 8. | UNIFSKIL | 9 8 7 6 5 4 3 2 1 2 3 4 5 6 7 8 9 | FLEXSKIL |
| 9. | SERVICES | 9 8 7 6 5 4 3 2 1 2 3 4 5 6 7 8 9 | SKILLS |

10. **Do you prefer uniform or flexible development policies across services?**
11. **Do you prefer uniform or flexible development policies across skills?**

12. **Do you feel more strongly about this issue across services or across skills?**

1=EQUAL 3=MODERATE 5=STRONG 7=VERY STRONG 9=EXTREME

| DEVELOPMENT | | | | | | | | | | | | | | | | | | | |
|---|---|---|---|---|---|---|---|---|---|---|---|---|---|---|---|---|---|---|---|
| 10. | UNIFSVC | 9 | 8 | 7 | 6 | 5 | 4 | 3 | 2 | 1 | 2 | 3 | 4 | 5 | 6 | 7 | 8 | 9 | FLEXSVC |
| 11. | UNIFSKIL | 9 | 8 | 7 | 6 | 5 | 4 | 3 | 2 | 1 | 2 | 3 | 4 | 5 | 6 | 7 | 8 | 9 | FLEXSKIL |
| 12. | SERVICES | 9 | 8 | 7 | 6 | 5 | 4 | 3 | 2 | 1 | 2 | 3 | 4 | 5 | 6 | 7 | 8 | 9 | SKILLS |

13. **Do you prefer uniform or flexible promotion policies across services?**

14. **Do you prefer uniform or flexible promotion policies across skills?**

15. **Do you feel more strongly about this issue across services or across skills?**

1=EQUAL 3=MODERATE 5=STRONG 7=VERY STRONG 9=EXTREME

| PROMOTION | | | | | | | | | | | | | | | | | | | |
|---|---|---|---|---|---|---|---|---|---|---|---|---|---|---|---|---|---|---|---|
| 13. | UNIFSVC | 9 | 8 | 7 | 6 | 5 | 4 | 3 | 2 | 1 | 2 | 3 | 4 | 5 | 6 | 7 | 8 | 9 | FLEXSVC |
| 14. | UNIFSKIL | 9 | 8 | 7 | 6 | 5 | 4 | 3 | 2 | 1 | 2 | 3 | 4 | 5 | 6 | 7 | 8 | 9 | FLEXSKILL |
| 15. | SERVICES | 9 | 8 | 7 | 6 | 5 | 4 | 3 | 2 | 1 | 2 | 3 | 4 | 5 | 6 | 7 | 8 | 9 | SKILLS |

16. **Do you prefer uniform or flexible transition policies across services?**

17. **Do you prefer uniform or flexible transition policies across skills?**

18. **Do you feel more strongly about this issue across services or across skills?**

1=EQUAL 3=MODERATE 5=STRONG 7=VERY STRONG 9=EXTREME

| TRANSITION | | | | | | | | | | | | | | | | | | | |
|---|---|---|---|---|---|---|---|---|---|---|---|---|---|---|---|---|---|---|---|
| 16. | UNIFSVC | 9 | 8 | 7 | 6 | 5 | 4 | 3 | 2 | 1 | 2 | 3 | 4 | 5 | 6 | 7 | 8 | 9 | FLEXSVC |
| 17. | UNIFSKIL | 9 | 8 | 7 | 6 | 5 | 4 | 3 | 2 | 1 | 2 | 3 | 4 | 5 | 6 | 7 | 8 | 9 | FLEXSKILL |
| 18. | SERVICES | 9 | 8 | 7 | 6 | 5 | 4 | 3 | 2 | 1 | 2 | 3 | 4 | 5 | 6 | 7 | 8 | 9 | SKILLS |

**Our next step will be to tabulate all the responses. We will provide you with your responses as well as those of the entire group. We will not identify individual respondents except to themselves. We are recommending that the views of the senior group about goals, objectives, and management principles be discussed among yourselves at the first senior meeting to discern where commonalty or disagreement might exist. Thank you again for your time and assistance.**

Appendix B

# EXAMPLE OF DETERMINING AN ASPECT

Chapter Five described the generic process of evaluating alternatives against objectives. This appendix provides an actual example from the study.[1] It details the process used for evaluating the alternatives for one aspect—the point at which line officers should enter a career. The three alternatives considered under this personnel management aspect are

A. Lateral entry from civilian life;

B. Lateral entry from the reserves or with prior service; and

C. Entry at year 0, no prior experience (the current practice for line officers).

For convenience, we refer to these alternatives as A, B, and C, respectively.

First we need to select a scale to measure how well each entry point alternative keeps costs reasonable. The measurement scale must be relevant to both the objective and the aspect. This objective relates to costs, and we logically start with a dollar scale with end points of higher costs and lower costs. The costs we are interested in also need to relate specifically to the entry point. Annualized accession and initial training cost data are available from the previous study.[2] We

---

[1]Our study contains over 800 evaluations of the type shown in this appendix. The documentation is too voluminous to include here but is available from the authors.

[2]See Harry Thie and Roger Brown, *Future Career Management Systems for U.S. Military Officers*, MR-470-OSD, p. 180.

can cost the status quo (alternative C) at $21,200 per line officer accession per year.[3] Accession and training costs, also available from the previous study, of $17,200 and $19,400 can be placed on the scale for alternatives A and B. These costs are plotted on the measurement scale in Figure B.1.

From the above, we can already rank the alternatives for the single objective of "keep costs reasonable," but we do not yet know how well each alternative's performance in "keep costs reasonable" compares with its performance on the other ten objectives. We could calculate, relative to alternative C, a percentage decrease (or increase) in costs for each of the alternatives A and B.[4] If we assigned values to the end points of the above measurement scale, we could calculate where each alternative lay on the scale as a percentage of the total length of the line. All of our answers would lie between 0 and 1. We would have *normalized* the evaluation, and we could assign a score to each alternative. This approach provides a way to

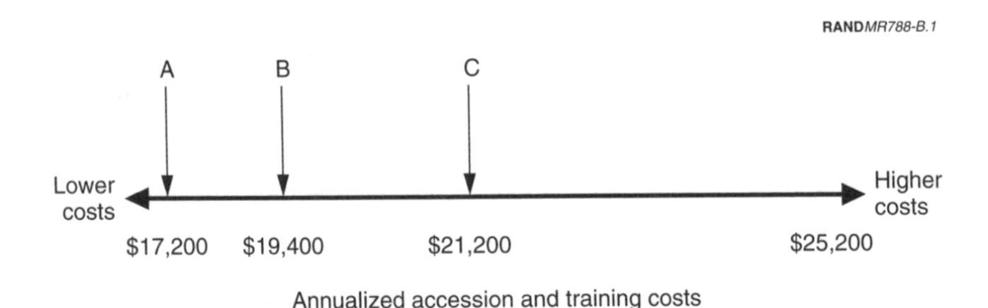


**Figure B.1—Measurement Scale with Values for Assessing Entry Point Alternatives Against the "Keep Costs Reasonable" Objective**

---

[3]These are costs in FY 1993 dollars and include the variable cost of entry from the different accession sources and initial skill training courses. This total cost is then amortized over the length of an expected career to produce the annualized cost. See MR-470-OSD for a more complete derivation of these costs.

[4]For alternative B, (21,200 – 19,400)/21,200 = 8.49% reduction in costs compared to alternative C; alternative A is an 18.87% lowering of costs compared to alternative C.

compare apples with oranges by normalizing to compare the achievement of different measures.[5]

To add the normalizing function to the above figure, we add a y-axis with values ranging from 0 to 1. We also add end point values to specify a range along the x-axis ranging from $17,200 to $25,200.[6] We determine the end points as follows. The current cost of acquiring an officer is $21,200. Alternative A represents almost a 19 percent savings, is a feasible full accomplishment of the objective, and becomes one end point. To determine the other end point, we move an equal amount in the opposite direction, that is, an almost 20 percent increase in costs, which policymakers would find unacceptable and thus fails to accomplish the objective. When we draw a line connecting the "$25,200" end point on the x-axis with the "1" end point on the y-axis, we can determine the normalized score for any accession and training cost value. This is depicted in Figure B.2.

Note from Figure B.2 that, regardless of the scales used, the scores for these alternatives will always reflect the same ranking on this objective. It should also be noted that equal changes in costs have equal value regardless of where those changes occur. Thus, the line (connecting the end points) represents equal policy implications for rising or decreasing costs. However, this assumption runs counter to what we would expect from policymakers. They have less concern about small cost differences around the status quo and more about differences at the extremes. To reflect better these concerns about how costs should be kept reasonable, we formulated a single-dimensional value function for this objective. Single-value functions define the relative importance of increments of change.[7] This function is displayed in Figure B.3.

---

[5]Thus, the percentage changes in costs associated with the "keep costs reasonable" objective can be compared with, for example, the changes in average career length associated with meeting active duty experience needs.

[6]The determination of end point values is an important step in properly scaling the evaluation score. The range of values must be relevant to the alternatives.

[7]As mentioned earlier, we adopt "single-value functions" as a shorthand version of single-dimensional functions. See Kirkwood, 1995, pp. 4-7 to 4-14. Value functions are important because once this function is determined, it is used in the evaluation of all alternatives for that specific objective. There is a single-dimensional value function for each of the 11 objectives. These functions were validated with the working group and separately with a member of the senior group.

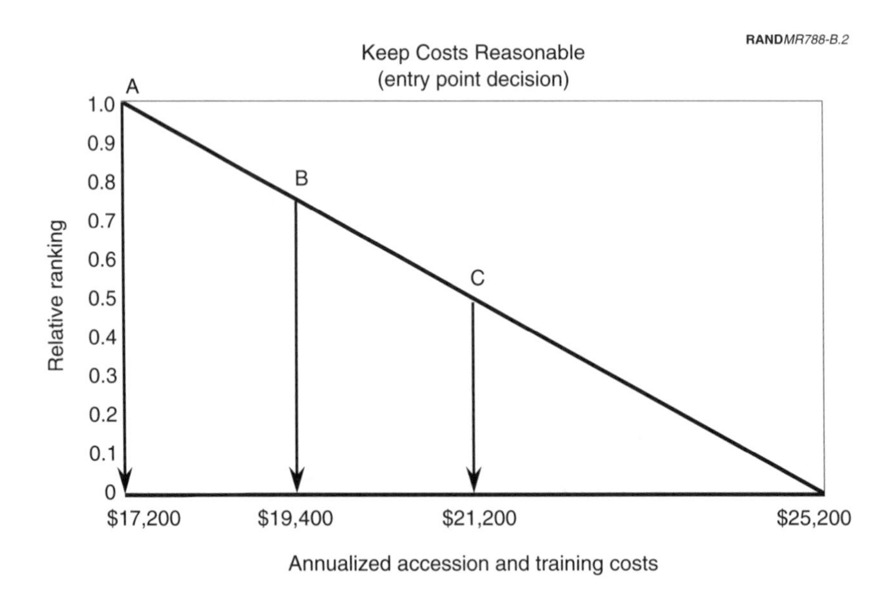


**Figure B.2—Example of Normalized Evaluation Scoring**

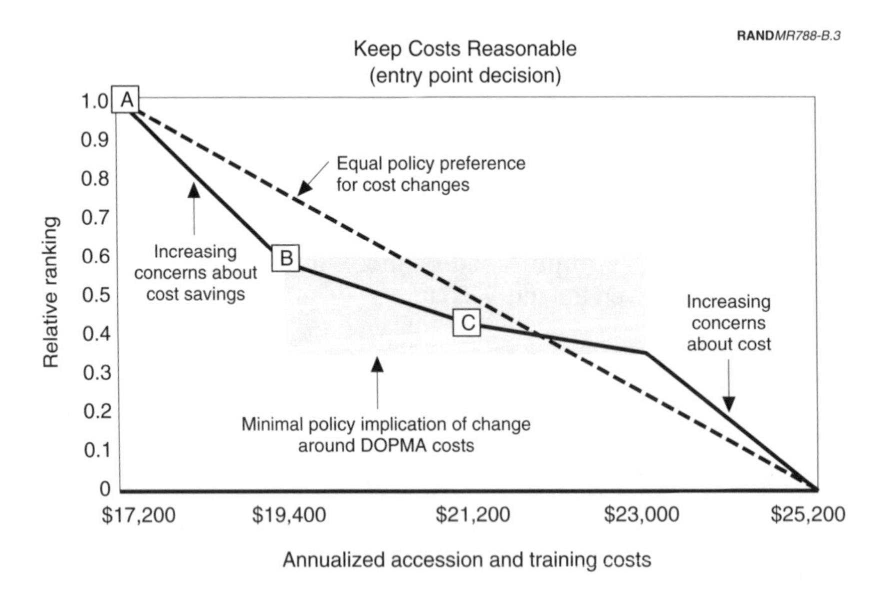


**Figure B.3—Example of Incorporating Preferences Within an Objective into the Evaluation Process**

Figure B.3 shows a clear preference for costs below $19,400 and a clear aversion for costs above $23,000. Between the two (within the shaded area), while the decreased costs are always preferred, the level of concern is significantly less. From the figure, we can calculate new value scores for each of our alternatives; for example, alternative C would now have a value score of 0.43. The status quo is no longer at the midpoint on the normalized scale; and the lower value (0.43) reflects the policymakers' dissatisfaction with maintaining status quo costs and the preference for cost savings. Alternative B falls just on the edge of the shaded area and represents a cost saving that, though desirable, has less significance to policymakers than the level of cost savings attainable through alternative A. The insight we gain from using the above value function is twofold: alternative A achieves savings that are of the highest order of significance; and while alternative B is arithmetically nearly midway between the other two, it would probably be perceived by policymakers as being closer to the status quo than to alternative A.

This example also illustrates an opportunity to simplify the evaluation process. Given that these single-value functions[8] do not change the ranking of alternatives, the evaluation score could be represented by discrete rather than calculated values along the continuous function. In other words, in Figure B.3, we can simply have five positions—significantly increasing costs, minimally increasing costs, status quo, minimally decreasing costs, and significantly decreasing costs. The corresponding scores for any evaluation within each of these ranges for this objective would be 0.0, 0.37, 0.43, 0.57, and 1.0. Rather than evaluate each alternative on its calculated score, we can assign a relative value score that evaluates the alternatives by assigning the score for relative position on the value function. For our example, the value scores would thus be 1.0 for alternative A; 0.57 for alternative B, and 0.43 for alternative C. Note that the rankings have been maintained, but that alternative A attains a higher relative score, reflecting its significant cost savings.

[8]As long as the function continuously increases, the ranking will be maintained. All of the single-value dimensional functions we developed continuously increased across the evaluation range.

We now have one decision in our example—lateral entry from civilian life has been evaluated as achieving the highest score in meeting the "keep costs reasonable" objective. Figure B.4 shows this.

However, this represents but one decision point in the evaluation of the three alternatives. We now need to evaluate in a similar manner each of the other ten objectives. When complete, we will have value scores and rankings for how well each alternative achieves each objective. But unless one alternative consistently scores high for every objective, we will have to resolve which alternative scores highest across all the objectives.

Several mathematical approaches can assist in this process. One could find the average score for each alternative across all of the objectives[9] and choose the highest. This approach assumes that each

RANDMR788-B.4

**Line Officers**

| Entry Point Alternative | Objective |
|---|---|
| | Keep costs reasonable |
| Year 0 | .43 |
| Lateral from reserves or prior service | .57 |
| Lateral from civilian life | 1.0 |

**Figure B.4—Value Scores for Entry Point Aspect Evaluation for "Keep Costs Reasonable" Objective**

[9]One would sum the individual value scores for an alternative and divide by 11. This would represent an average or arithmetic mean.

objective is of equal importance or weight. However, from Chapter Four we know the policymakers have varying preferences for the objectives, and we can use this information to evaluate how well an alternative satisfies all of the objectives. By so doing, we can determine which alternative best achieves all objectives in a manner that reflects the preferences of the policymaking group.

To evaluate the alternatives across all objectives, the priority or weighting of the objective is applied to the value scores. The application of the weighting does not change the relative order of the alternatives within individual objective evaluations but permits the alternative to be evaluated across all the objectives. After the weights have been applied, the alternatives are ranked. Figure B.5 displays how the weight is applied.

The objective weight represents the preferences for objectives by the policymaking group, as discussed in Chapter Four.

RANDMR788-B.5

**Line Officers**

| | Objective |
|---|---|
| | Keep costs reasonable |
| Entry Point Alternative | Weight (%) = 19 |
| Year 0 | .43 x .19 = .08 |
| Lateral from reserves or prior service | .57 x .19 = .11 |
| Lateral from civilian life | 1.0 x .19 = .19 |

**Figure B.5—Applying Objective Weight to Value Scores**

The same process of selecting scales, developing value scores, and applying objective weights is then done for each objective. When all have been completed, an overall score can be calculated by summing the individual objective scores for each alternative as depicted in Figure B.6. The last column on the right represents the sum of those scores, with Year 0 achieving the highest score.

Figure B.6 depicts the final decision for this set of alternatives. Note how the alternatives evaluate differently for each objective. Though lateral entry from civilian life scores high in achieving the "keep costs reasonable" objective, it does poorly in virtually all other objectives. It ranks last or tied for last in all but three of the remaining objectives; of the remaining three, it ties for first in all. Its final score is the lowest of the three alternatives.

RANDMR788-B.6

Line Officers

| Entry Point Alternative | Objective | | | | | | | | | | | Decision score |
|---|---|---|---|---|---|---|---|---|---|---|---|---|
| | Keep costs reasonable | Provide career satisfaction | Emphasize cadre with military culture | Meet active experience needs | Meet active skill needs | Inculcate culture prior to or at entry | Provide high opportunity to serve | Provide career opportunity | Meet reserve needs | Meet active grade needs | Be compatible with civilian careers | |
| | Weight (%) | | | | | | | | | | | |
| | 19 | 13 | 13 | 13 | 12 | 8 | 7 | 6 | 5 | 3 | 2 | |
| Year 0 | .08 | .10 | .09 | .06 | .09 | .05 | .05 | .03 | .02 | .02 | 0 | .59 |
| Lateral from reserves or prior service | .11 | .07 | .11 | .06 | .09 | .05 | 0 | .01 | 0 | .02 | .02 | .54 |
| Lateral from civilian life | .19 | 0 | 0 | 0 | 0 | 0 | .05 | .01 | 0 | .02 | .02 | .29 |

**Figure B.6—Final Decision Table for Entry Point Alternatives**

Appendix C

# SINGLE-VALUE FUNCTIONS

This appendix contains the single-value functions for each of the 11 objectives. The discrete value function provides the relative value of the evaluation increments. Tables C.1–C.15 include the change from the status quo, the values of each increment, and the cumulative values.

We use two procedures for determining single-dimensional value functions. (See Kirkwood, 1995.) One procedure results in a function that is made up of segments of straight lines that are joined together into a piecewise linear function. The other procedure uses a specific mathematical form, exponential, for the value function. Piecewise linear functions are typically used when there are a small number of different scoring levels. (See, for example, Table C.1.) Exponential functions are used when there are a large number (essentially an infinite number) of different levels between the end points. (See, for example, Table C.4.) These functions monotonically increase or decrease. (See Kirkwood and Sarin, 1980.)

Table C.1 provides the discrete value function for the objective "meet active skill needs." The end points, which differ for many of the objectives, are –2 (significantly detracts from the status quo) and 2 (significantly improves the status quo). The increment between significantly detracts from the status quo and detracts is valued at .538, or 54 percent. The increment between detracts and no difference is valued at .231. Thus, the cumulative value of no difference is 77 percent.

Table C.4 is an example of the second kind of value function for continuous variables. This table permits the assessment of specific data

**Table C.1**

**Meet Active Skill Needs**

| Change from Status Quo | Value | Cumulative Value (%) |
|---|---|---|
| –2=Significantly detracts | 0.000 | 0 |
| –1=Detracts | 0.538 | 54 |
| 0=No difference | 0.231 | 77 |
| 1=Improves | 0.077 | 85 |
| 2=Significantly improves | 0.154 | 100 |
| Sum of increments | 1.000 | |

**Table C.2**

**Meet Active Grade Needs**

| Change from Status Quo | Value | Cumulative Value (%) |
|---|---|---|
| 1=Detracts | 0.000 | 0 |
| 0=No difference | 0.500 | 50 |
| 1=Improves | 0.250 | 75 |
| 2=Significantly improves | 0.250 | 100 |
| Sum of increments | 1.000 | |

**Table C.3**

**Meet Active Experience Needs**

| Change from Status Quo | Value | Cumulative Value (%) |
|---|---|---|
| –2=Significantly detracts | 0.000 | 0 |
| –1=Detracts | 0.250 | 25 |
| 0=No difference | 0.250 | 50 |
| 1=Improves | 0.250 | 75 |
| 2=Significantly improves | 0.250 | 100 |

from earlier studies with a large number of possible scores. In this case, the data analyzed are the average years of experience of the officer force, and the single-value function is exponential.[1]

---

[1]We used Excel spreadsheets as suggested by Kirkwood (1995) to derive and use these functions.

**Table C.4**

**Meet Active Experience Needs (data-based)**

| Change from Status Quo (%) | Value | Cumulative Value (%) |
|---|---|---|
| –5 | 0.000 | 0 |
| –4 | 0.284 | 28 |
| –3 | 0.203 | 49 |
| –2 | 0.146 | 63 |
| –1 | 0.104 | 74 |
| 0 | 0.075 | 81 |
| 1 | 0.054 | 87 |
| 2 | 0.038 | 90 |
| 3 | 0.028 | 93 |
| 4 | 0.020 | 95 |
| 5 | 0.014 | 97 |
| 6 | 0.010 | 98 |
| 7 | 0.007 | 98 |
| 8 | 0.005 | 99 |
| 9 | 0.004 | 99 |
| 10 | 0.003 | 99 |
| 11 | 0.002 | 100 |
| 12 | 0.001 | 100 |
| 13 | 0.001 | 100 |
| 14 | 0.001 | 100 |
| 15 | 0.001 | 100 |

NOTE: The evaluation measure used here has a large number of possible scores. Integer values for some of these scores are shown in the table. This single-dimensional-value function monotonically increases (larger amounts of active duty experience are preferred to smaller amounts). We follow procedures outlined in Kirkwood (1995) and Kirkwood and Sarin (1980) and use spreadsheets to display particular values (shown above) and graph (not shown) the following exponential function ($\rho = .03$):

$$v(x) = \frac{e^{\frac{-(x-(-5)\%)}{\rho}} - 1}{e^{\frac{-(15\%-(-5)\%)}{\rho}} - 1}$$

**Table C.5**

**Meet Reserve Needs**

| Change from Status Quo | Value | Cumulative Value (%) |
|---|---|---|
| –2=Significantly detracts | 0.000 | 0 |
| –1=Detracts | 0.071 | 7 |
| 0=No difference | 0.214 | 29 |
| 1=Improves | 0.214 | 50 |
| 2=Significantly improves | 0.500 | 100 |
| Sum of increments | 1.000 | |

**Table C.6**

**Keep Costs Reasonable**

| Change from Status Quo Costs | Value | Cumulative Value (%) |
|---|---|---|
| –2=Adds a lot of cost | 0.000 | 0 |
| –1=Adds some cost | 0.357 | 36 |
| 0=No difference | 0.071 | 43 |
| 1=Reduces costs | 0.143 | 57 |
| 2=Significantly reduces costs | 0.429 | 100 |
| Sum of increments | 1.000 | |

**Table C.7**

**Keep Costs Reasonable (for specific costing data)**

| Change from Status Quo (%) | Value | Cumulative Value (%) |
|---|---|---|
| 8 | 0.000 | 0 |
| 7 | 0.036 | 4 |
| 6 | 0.038 | 7 |
| 5 | 0.040 | 11 |
| 4 | 0.043 | 16 |
| 3 | 0.045 | 20 |
| 1 | 0.048 | 25 |
| 0 | 0.051 | 30 |
| –1 | 0.054 | 35 |
| –2 | 0.057 | 41 |
| –3 | 0.060 | 47 |
| –4 | 0.063 | 53 |
| –5 | 0.067 | 60 |
| –6 | 0.071 | 67 |
| –8 | 0.075 | 75 |
| –9 | 0.079 | 83 |
| –10 | 0.084 | 91 |
| –11 | 0.089 | 100 |

NOTE: The evaluation measure used here has a large number of possible scores. Integer values for some of the possible scores are shown in the table. This single-dimensional-value function is monotonically increasing (larger amounts of cost savings are preferred to smaller amounts). We followed procedures outlined in Kirkwood (1995) and Kirkwood and Sarin (1980) and used spreadsheets to display particular values (shown above) and graph (not shown) the following exponential function ($\rho = .2$):

$$v(x) = \frac{e^{\frac{-(x-8\%)}{\rho}} - 1}{e^{\frac{-(-11\%-8\%)}{\rho}} - 1}$$

**Table C.8**

**Provide High Opportunity to Serve**

| Change from Status Quo | Value | Cumulative Value (%) |
|---|---|---|
| –1=Detracts | 0.000 | 0 |
| 0=No difference | 0.700 | 70 |
| 1=Improves | 0.200 | 90 |
| 2=Significantly improves | 0.100 | 100 |
| Sum of increments | 1.000 | |

**Table C.9**

**Provide High Opportunity to Serve (for specific accession data)**

| Change from Status Quo (%) | Value | Cumulative Value (%) |
|---|---|---|
| –53 | 0.000 | 0 |
| –47 | 0.114 | 11 |
| –42 | 0.102 | 22 |
| –37 | 0.092 | 31 |
| –32 | 0.083 | 39 |
| –26 | 0.075 | 47 |
| –21 | 0.067 | 53 |
| –16 | 0.060 | 59 |
| –11 | 0.054 | 65 |
| –5 | 0.049 | 70 |
| 0 | 0.044 | 74 |
| 5 | 0.040 | 78 |
| 11 | 0.036 | 82 |
| 16 | 0.032 | 85 |
| 21 | 0.029 | 88 |
| 26 | 0.026 | 90 |
| 32 | 0.023 | 93 |
| 37 | 0.021 | 95 |
| 42 | 0.019 | 97 |
| 47 | 0.017 | 98 |
| 53 | 0.015 | 100 |

NOTE: The evaluation measure used here has a large number of possible scores. Integer values for some of the possible scores are shown in the table. This single dimensional value function is monotonically increasing (larger amounts of cost savings are preferred to smaller amounts). We followed procedures outlined in Kirkwood (1995) and Kirkwood and Sarin (1980) and used spreadsheets to display particular values (shown above) and graph (not shown) the following exponential function ($\rho = .5$):

$$v(x) = \frac{e^{\frac{-(x-53\%)}{\rho}} - 1}{e^{\frac{-(-53\%-53\%)}{\rho}} - 1}$$

## Table C.10

## Emphasize Military Culture Prior to Entry

| Change from Status Quo | Value | Cumulative Value (%) |
|---|---|---|
| –2=Significantly detracts | 0.000 | 0 |
| –1=Detracts | 0.556 | 56 |
| 0=No difference | 0.111 | 67 |
| 1=Improves | 0.111 | 78 |
| 2=Significantly improves | 0.222 | 100 |
| Sum of increments | 1.000 | |

## Table C.11

## Value Cadre of Officers with Military Culture

| Change from Status Quo | Value | Cumulative Value (%) |
|---|---|---|
| –2=Significantly detracts | 0.000 | 0 |
| –1=Detracts | 0.455 | 45 |
| 0=No difference | 0.182 | 64 |
| 1=Improves | 0.182 | 82 |
| 2=Significantly improves | 0.182 | 100 |
| Sum of increments | 1.000 | |

## Table C.12

## Provide Career Satisfaction

| Change from Status Quo | Value | Cumulative Value (%) |
|---|---|---|
| –2=Significantly detracts | 0.000 | 0 |
| –1=Detracts | 0.538 | 54 |
| 0=No difference | 0.231 | 77 |
| 1=Improves | 0.154 | 92 |
| 2=Significantly improves | 0.077 | 100 |
| Sum of increments | 1.000 | |

**Table C.13**

**Provide Career Opportunity**

| Change from Status Quo | Value | Cumulative Value (%) |
|---|---|---|
| –2=Significantly detracts | 0.000 | 0 |
| –1=Detracts | 0.091 | 9 |
| 0=No difference | 0.455 | 55 |
| 1=Improves | 0.364 | 91 |
| 2=Significantly improves | 0.091 | 100 |
| Sum of increments | 1.000 | |

**Table C.14**

**Provide Career Opportunity (data-based)**

| Change from Status Quo (12.7 years of expected career length) | Value | Cumulative Value (%) |
|---|---|---|
| –1.7 | 0.000 | 0 |
| –1.2 | 0.320 | 32 |
| –.7 | 0.218 | 54 |
| –.2 | 0.148 | 69 |
| +.3 | 0.101 | 79 |
| +.8 | 0.069 | 86 |
| +1.3 | 0.047 | 90 |
| +1.8 | 0.032 | 93 |
| +2.3 | 0.022 | 96 |
| +2.8 | 0.015 | 97 |
| +3.3 | 0.010 | 98 |
| +3.8 | 0.007 | 99 |
| +4.3 | 0.005 | 99 |
| +4.8 | 0.003 | 100 |
| +5.3 | 0.002 | 100 |
| +5.8 | 0.001 | 100 |
| +6.3 | 0.001 | 100 |

NOTE: The evaluation measure used here has a large number of possible scores and some of the possible scores are shown in the table. This single-dimensional-value function is monotonically increasing (larger amounts of career opportunity are preferred to smaller amounts). We followed procedures outlined in Kirkwood (1995) and Kirkwood and Sarin (1980) and used spreadsheets to display particular values (shown above) and graph (not shown) the following exponential function ($\rho = 1.3$):

$$v(x) = \frac{e^{\frac{-(x-11)}{\rho}} - 1}{e^{\frac{-(19-11)}{\rho}} - 1}$$

**Table C.15**

**Maintain Compatibility with Civilian Sector Careers**

| Change from Status Quo | Value | Cumulative Value (%) |
|---|---|---|
| –2=Not compatible | 0.000 | 0 |
| –1=Less compatible | 0.143 | 14 |
| 0=Compatible | 0.857 | 100 |
| Sum of increments | 1.000 | |

Appendix D

# SENSITIVITY ANALYSIS

This appendix discuses the sensitivity analysis we conducted to illustrate the effects of changing objective weights.

## KEEP COSTS REASONABLE

The highest group objective weight used in developing the preferred OCMS was "keep costs reasonable" (19 percent), and the preferred system reflects this emphasis. Three members of the decisionmaker group ranked "keep costs reasonable" as their top priority. The decisionmaker with the highest weight (59 percent) for reasonable costs had "meet active experience needs" as his second priority (13 percent). For our sensitivity analysis, we used his set of objective weights to determine which alternatives were affected by objective weights reflecting increased emphasis on "keep costs reasonable." The following alternatives differed from those in the preferred OCMS:

### Accessing

- The career system would become an open system with maximum dependence on new officers bringing needed experience and skills acquired in civilian life, entering at grades appropriate to their civilian background.
- While enlisted service would continue to be a desired alternative, there would be increased emphasis on accession programs requiring no acculturation.

**Developing**[1]

- Less military and civilian education

**Promoting**

- A promotion zone based on time in service[2]

**Transitioning**

- No vesting.

## PROVIDE CAREER SATISFACTION

This objective had the second highest weight (13 percent) in the overall averages and was the top priority of two of the policymakers. The decisionmaker's weights we used for our sensitivity analysis had "provide career satisfaction" at 49 percent with "provide career opportunity" having the second highest weight at 14 percent.

**Accessing**

- A shorter payback period roughly equal to the time spent in pre-entry education and training

**Developing**

- Slightly longer assignments (up to one year longer than the current practice)

---

[1]From a pure cost standpoint, officers not selected for career status would be separated. However, for this particular set of objective weights, some officers who do not select for career status would be required to transfer to another skill group. This is an example of how the choice of an alternative is affected by the weights of all of the alternatives. In this case, the weight associated with meeting experience needs was sufficient to change the choice.

[2]This choice was driven as much by the weight given to meeting experience as to keeping costs reasonable.

**Promoting**

- An officer would be eligible for promotion as long as he is on active duty (open zone)[3]

**Transitioning**

- Intermediate tenure, i.e., a promise of continued service given attainment of a certain grade.

## EMPHASIZE CADRE WITH MILITARY CULTURE

An objective weight of 13 percent was used in the preferred system. The decisionmaker with this as his highest weight (31 percent) had "provide career opportunity" second (11 percent), which we used for our sensitivity analysis.

**Accessing**

- A very short payback period for pre-entry education and training (0.5:1)

**Developing**

- (No change from the alternatives in the preferred OCMS)

**Promoting**

- An officer would be eligible for promotion as long as he is on active duty (open zone)

**Transitioning**

- Intermediate tenure.

## MEET ACTIVE EXPERIENCE NEEDS

An objective weight of 12 percent was used in the preferred system. The decisionmaker with this as his highest weight (29 percent) had

[3]If only career satisfaction was being considered, time in grade would be used to determine the promotion zone.

"emphasize cadre with military culture" second (20 percent), which we used for our sensitivity analysis.

**Accessing**

- (No change from the alternatives in the preferred OCMS)

**Developing**

- Slightly longer assignments (up to one year longer than the current practice)

**Promoting**

- A promotion zone based on time in service
- An officer would be eligible for promotion as long as he is on active duty (open zone)

**Transitioning**

- No vesting
- Separation pay for involuntary separation
- Moderate turnover early in service.

## MEET ACTIVE SKILL NEEDS

An objective weight of 11 percent was used in the preferred system. The decisionmaker with this as his highest weight (40 percent) had "meet active experience needs" second (17 percent), which we used for our sensitivity analysis.

**Accessing**

- (No change from the alternatives in the preferred OCMS)

**Developing**

- (No change from the alternatives in the preferred OCMS)

**Promoting**

- A promotion zone based on time in service

- An officer would be eligible for promotion as long as he is on active duty (open zone)

**Transitioning**

- No vesting
- Moderate turnover early in service
- Length of career similar to today's.[4]

## INCULCATE CULTURE PRIOR TO OR AT ENTRY

An objective weight of 8 percent was used in the preferred system. The decisionmaker with this as his highest weight (48 percent) had "keep costs reasonable" second (20 percent).

**Accessing**

- While enlisted service would continue to be a desired alternative, there would be increased emphasis on accession programs that are of high intensity, educational, and of long duration

**Developing**

- Less military and civilian education

**Promoting**

- (No change from the alternatives in the preferred OCMS)

**Transitioning**

- (No change from the alternatives in the preferred OCMS).

---

[4]If only meeting skill needs was being considered, the preferred transition alternatives would be those having maximum career lengths of 30 or more years, a retirement annuity after 25 or more years, and low-to-moderate separation of officers in the early years.

## PROVIDE HIGH OPPORTUNITY TO SERVE

An objective weight of 7 percent was used in the preferred system. The decisionmaker with this as his highest weight (20 percent) had "provide career opportunity" second (15 percent).

**Accessing**

- A shorter payback period roughly equal to the time spent in pre-entry education and training
- A four-year period of initial tenure for those aspiring to longer careers

**Developing**

- (No change from the alternatives in the preferred OCMS)

**Promoting**

- (No change from the alternatives in the preferred OCMS)[5]

**Transitioning**

- Separation pay for involuntary separation.

---

[5]If only opportunity to serve was being considered, separation would be the preferred alternative for those officers who were not promoted.

Appendix E

# FEASIBILITY OF SUPPLY

The authors used multiobjective decision analysis to determine which alternative best satisfied the set of prioritized objectives for each of the aspects listed under the four career management functions. Each element of a future career management system was selected because it best satisfied the objectives as they were prioritized by the policymakers. However, the future career management system is a system in which the alternatives will interact with one another, rather than just a set of distinct parts. In other words, the ways in which officers enter, move through, and exit the system need to be internally consistent to be able to support the military manpower requirements for officers; if too many (or too few) enter or exit the system in the wrong places, the system will not support established requirements. Thus, we conducted some basic system feasibility modeling.

We used a system dynamics approach to simulate the interactions of the career alternatives proposed and to model the flow of officers through the career system. The modeling was based upon several sets of inputs. First, the services provided future requirements for officers by grade. We divided these total requirements into requirements for officers in each of the four skill groups, based upon the current division of skill groups for each grade. That is, if 67 percent of current Army O1s are line officers, then 67 percent of the Army O1s in the future requirements were also assumed to be line officers. The requirements are shown in Table E.1. We did not question whether these were the "right" requirements but simply determined if the career management system could support them.

**Table E.1**

**Officer Requirements**

| Service | O1–O3 | O4 | O5 | O6 | Total |
|---|---|---|---|---|---|
| Army | | | | | |
| Line | 23464 | 6966 | 3998 | 1443 | 35871 |
| Support | 6005 | 2433 | 1800 | 584 | 10822 |
| Specialist | 2267 | 1476 | 1001 | 372 | 5116 |
| Professional | 8843 | 4152 | 2477 | 1357 | 16829 |
| | 40579 | | | | 68638 |
| Navy | | | | | |
| Line | 11045 | 3676 | 3041 | 1758 | 19520 |
| Support | 5896 | 1948 | 1186 | 335 | 9365 |
| Specialist | 8919 | 2113 | 795 | 202 | 12029 |
| Professional | 6955 | 2844 | 1714 | 655 | 12168 |
| | 32815 | | | | 53082 |
| Marine Corps | | | | | |
| Line | 7204 | 2041 | 1235 | 574 | 11054 |
| Support | 2543 | 679 | 252 | 14 | 3488 |
| Specialist | 751 | 315 | 112 | 26 | 1204 |
| Professional | 235 | 122 | 35 | 8 | 400 |
| | 10733 | | | | 16146 |
| Air Force | | | | | |
| Line | 19617 | 5996 | 4660 | 1488 | 31761 |
| Support | 5449 | 1991 | 1519 | 716 | 9675 |
| Specialist | 10522 | 3569 | 2076 | 681 | 16848 |
| Professional | 8328 | 3896 | 2158 | 1037 | 15419 |

NOTE: Derivation of continuation rates was shown in Appendix G of MR-470-OSD. These year-to-year continuation rates for each cohort are service-specific and *jointly* reflect voluntary and involuntary losses.

The second key input to the model is service-specific continuation and loss rates. The rates were derived in MR-470-OSD and reflect the likely effect that specific policies would have on the rate and point at which officers leave the service.

Figure E.1 displays output from our modeling, and indicates that the officer system that results from the combination of alternatives is a coherent, feasible system with sufficient numbers of officers flowing through the system in a steady-state.

Besides indicating the feasibility of the future officer system, the model output also emphasized the degree to which the future requirements-based system would differ from service to service because of the different grade structures and requirements for line officers. Figures E.1 through E.4 display the model output for line officers in each of the four services. The legends include the proportion of line officers at each grade, and the annotations to the line graphs indicate a likely percentage of officers who will be promoted to each of the grades.[1] Thus, 92 percent of Navy O4s (see Figure E.2), who enter the promotion zone for consideration are promoted to O5s, based on the requirements and other assumptions (e.g., continuation rates) of the future system. This is considerably different from the 68 percent promotion rate for Army officers to the grade of O5 shown in Figure E.1. The difference is based upon the grade structures for line officers. Twenty percent of Navy line officers are O5s and O6s, compared to only 15 percent of Army officers. Thus, the Navy grade structure pulls officers through the system to satisfy their requirements at the higher grades.

---

[1]The percentage promoted is the percentage of all officers who enter the promotion zone for consideration. Because of the high-turnover-early characteristic of the objectives-based career system, proportionally fewer officers enter into longer careers. Potentially, this will increase grade-to-grade promotion opportunities given grade needs and continuation rates. Greater turnover early can increase promotion opportunity later, all other things being equal.

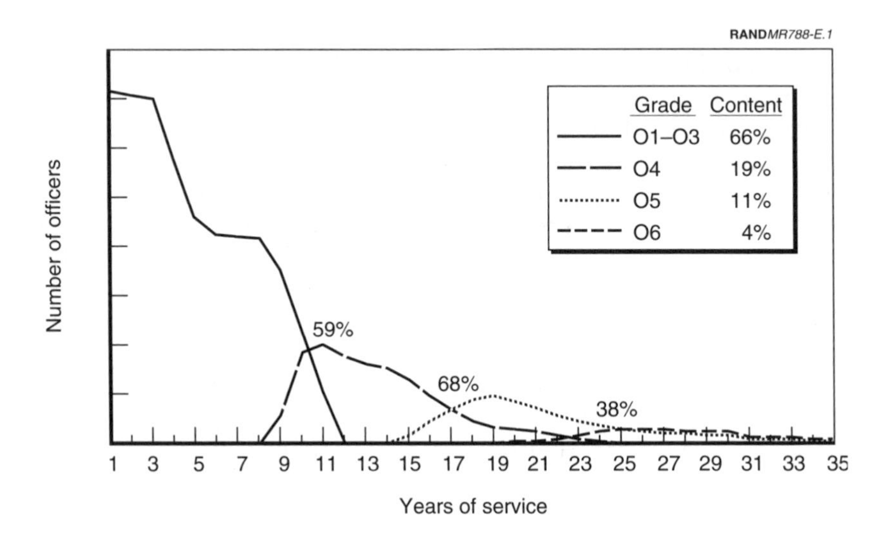


Figure E.1—Army Line Officer Flow and Promotions

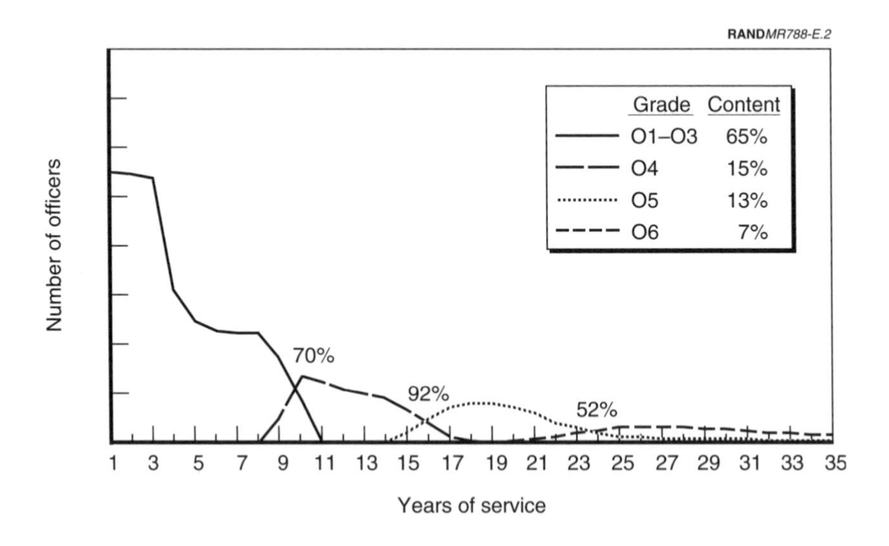


Figure E.2—Navy Line Officer Flow and Promotions

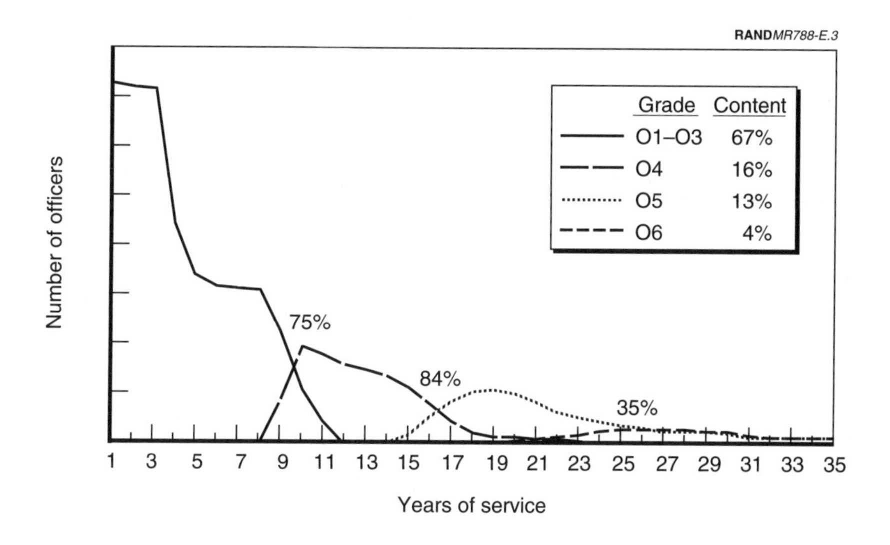


Figure E.3—Air Force Line Officer Flow and Promotions

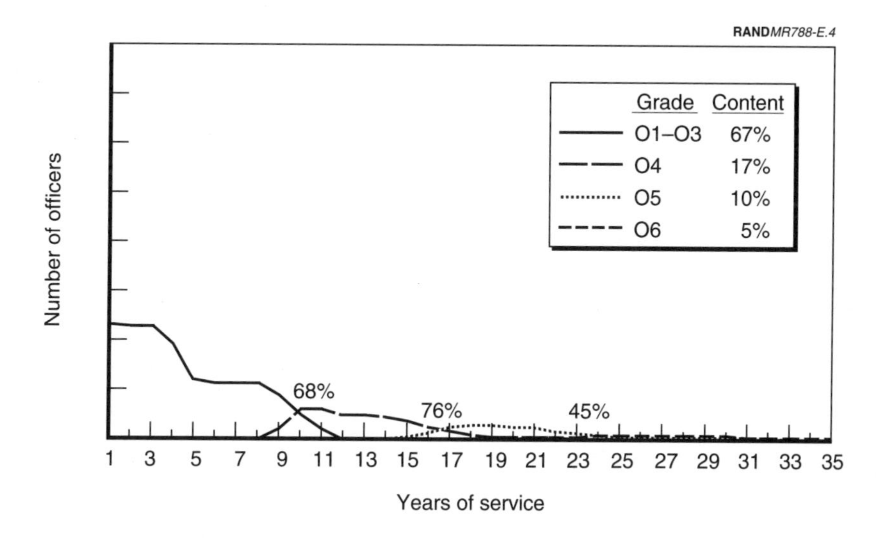


Figure E.4—Marine Corps Line Officer Flow and Promotions

Appendix F

# OTHER SKILL GROUPS

## OBJECTIVE WEIGHTS ACROSS SKILLS

As discussed in the context of sensitivity analysis in Chapter Six, differences in objective priorities can affect the resulting officer career management system. The objective weights used to determine the OCMS for line officers do not necessarily apply when considering the other skill groups. For example, inculcating the military culture before entry may be less important for managing military doctors than it is for line officers. Likewise, some might argue that skill requirements are more important for managing doctors than for line officers.

When the policymakers were interviewed, they were specifically asked for their preferences among objectives relative to military line officers, but opportunity for unstructured discussion of other officer skill groups was provided. The objective weights for the other skill groups were derived from the objective weights for line officers, as follows: First, the study team self-administered the interview for each of the skill groups. The proportionate differences between the study group's preferences for objectives for line officers and for each of the other skill groups were calculated. These proportionate differences were applied to the policymaking group's priorities for line officers to determine the priorities for each of the skill groups. Thus, the weights proffered here are not decisionmaker preferences, but rather are our assessment of how such preferences might change.

Table F.1 reflects the objective weights for each of the four officer skill groups.

**Table F.1**

**Officer Skill Group Prioritized Objectives (in percentage)**

| Objective | Line | Specialist | Support | Professional |
|---|---|---|---|---|
| Keep costs reasonable | 19 | 16 | 20 | 19 |
| Provide career satisfaction | 13 | 15 | 13 | 10 |
| Emphasize cadre with military culture | 13 | 13 | 8 | 3 |
| Meet active experience needs | 12 | 7 | 6 | 5 |
| Meet active skill needs | 11 | 20 | 13 | 26 |
| Inculcate culture prior to or at entry | 8 | 7 | 6 | 8 |
| Provide high opportunity to serve | 7 | 6 | 11 | 5 |
| Provide career opportunity | 6 | 9 | 7 | 10 |
| Meet reserve needs | 5 | 3 | 10 | 7 |
| Meet active grade needs | 3 | 3 | 2 | 3 |
| Be compatible with civilian careers | 2 | 1 | 4 | 5 |

A review of the changes in objectives across skill groups, as developed in this manner, validated both the approach used and the need for separate skill groups. Specialists possess skills unique to the military, and those skills can be taught only within the military, frequently at significant cost. Less emphasis on "keep costs reasonable," along with increased emphasis on "provide career satisfaction," are logical changes to the objectives. Support officers have skills that frequently have private sector counterparts; accordingly, their availability and utility to reserve forces as well as compatibility with the private sector would logically be expected to increase in comparison with line officers. Professionals, such as physicians, are frequently scarce, which is noted by the priority accorded to "meet active skill needs." Conversely, the need to "emphasize cadre with military culture" appears to be of less interest than with other skill groups.

The alternative evaluations were then conducted as described previously using quantitative and qualitative data applicable to each skill group.

## RESULTING CAREER MANAGEMENT SYSTEMS FOR OTHER SKILL GROUPS

In general, the career management systems that emerge from the multiobjective analysis for the other three skill groups are similar to the system for line officers. However, some differences emerge, and these are discussed below.

### Accessing Other Skill Groups

Officers of other skill groups would enter the career system differently. Specialist officers would enter laterally from other skill groups, as they currently do. Support and professional officers could enter the system laterally from civilian life, without an acculturation process before entry. The initial tenure period and the obligated service incurred for educational assistance would remain the same as for line officers.

### Promoting Other Skill Groups

All the skill groups would have the same nominal promotion system. In other words, they would be eligible for promotion based upon their time in grade, given a minimum time in service. The promotion zone would be long with selective promotion opportunity. If officers were not selected for promotion, they would be selectively continued. It is reasonable to assume, however, that the continuation rate for officers in the various skill groups could vary considerably.

### Developing Other Skill Groups

Developing officers of other skill groups differs by nonselection for career status. If a line officer is not selected for career status within that skill, he or she might be directed to migrate to another skill. They might, for example, become support officers. This would occur early in careers (before ten years of service) but not later. If support officers or professionals are not selected for career status within their skill, they will be separated.

### Transitioning Other Skill Groups

Officers in the other skill groups would transition out of the service in the same manner as line officers.

## UNIFORMITY AND FLEXIBILITY ACROSS SKILLS

The differences reflected in the objective weights and career systems for the other skill groups are consistent with the interview responses of the decisionmaker group regarding uniformity and flexibility across skills. As shown in Figure F.1, the policymaking group heavily favored flexibility in management policies across skills.

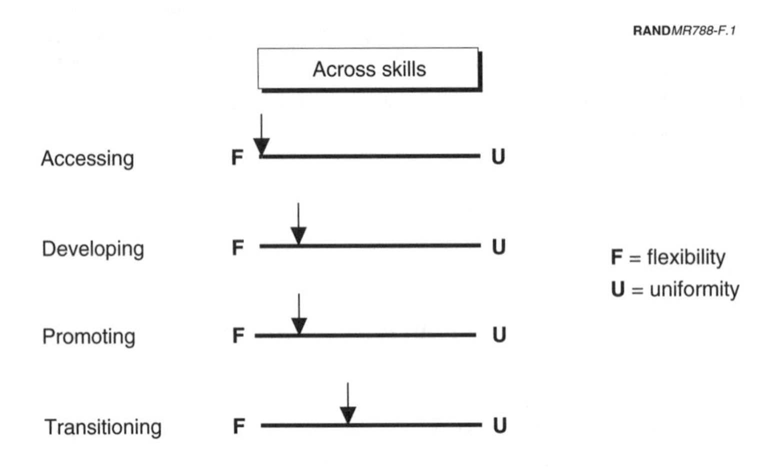


**Figure F.1—Preference for Flexibility and Uniformity Across Skills**

# REFERENCES

Adelman, Leonard, Paul J. Sticha, and Michael L. Donnell, "The Role of Task Properties in Determining the Relative Effectiveness of Multiattribute Weighting Techniques," *Organizational Behavior and Human Performance*, Vol. 13, 1975, pp. 31–45.

Asch, Beth J., and John T. Warner, "Should the Military Retirement System Be Reformed?" in J. Eric Fredland, Curtis L. Gilroy, Roger D. Little, and W. S. Sellman (eds.), *Professionals on the Front Line: Two Decades of the All-Volunteer Force*, Washington, DC: Brassey's, 1996.

Asch, Beth J., and John T. Warner, *A Policy Analysis of Alternative Military Retirement Systems*, MR-465-OSD, Santa Monica, CA: RAND, 1994.

Barron, Hutton, and Charles P. Schmidt, "Sensitivity Analysis of Additive Multiattribute Value Models," *Operations Research*, Vol. 36, No. 1, January–February 1988, pp. 122–127.

Borcherding, Katrin, Thomas Eppel, and Detlof von Winterfeldt, *Comparison of Weighting Judgments in Multiattribute Utility Measurement*, paper, October 1989.

Borcherding, Katrin, and Detlof von Winterfeldt, "The Effect of Varying Value Trees on Multiattribute Evaluations," *Acta Psychologica*, Vol. 68, 1988, pp. 153–170.

Brown, Rex V., "The State of the Art of Decision Analysis: A Personal Perspective," *Interfaces*, Vol. 22, No. 6, November-December 1992, pp. 5–14.

Buede, Dennis M., and Terry A. Bresnick, "Applications of Decision Analysis to the Military Systems Acquisition Process," *Interfaces*, Vol. 22, No. 6, November-December 1992, pp. 110–125.

Buede, Dennis M., "Structuring Value Attributes," *Interfaces*, Vol. 16, March-April 1986, pp. 52–62.

Carlsson, Christer, and Pirkko Walden, "AHP in Political Group Decisions: A Study in the Art of Possibilities," *Interfaces*, Vol. 25, July–August 1995, pp. 14–29.

Dyer, James S., "Remarks on the Analytic Hierarchy Process," *Management Science*, Vol. 36, No. 3, March 1990, pp. 249–275.

Dyer, James S., "A Clarification of 'Remarks on the Analytic Hierarchy Process'," *Management Science*, Vol. 36, No. 3, March 1990, pp. 274–275.

Gregory, Robin, and Keeney, Ralph L., "Creating Policy Alternatives Using Stakeholder Values," *Management Science*, Vol. 40, No. 8, August 1994, pp. 1035–1048.

Hall, Douglass T., *Careers in Organizations*, Pacific Palisades, CA: Goodyear Publishing Company, Inc., 1976.

Harsanyi, John C., "Cardinal Welfare, Individualist Ethics, and Interpersonal Comparisons of Utility," *Journal of Political Economy*, Vol. 63, 1955, pp. 309–321.

Keeney, Ralph L., *Manpower Planning and Personnel Management Models Based on Utility Theory*, San Francisco, CA: Woodward-Clyde Consultants, 1980.

Keeney, Ralph L., "Building Models of Values," *European Journal of Operations Research*, Vol. 37, 1988, pp. 149–157.

Keeney, Ralph L., "Using Values in Operations Research," *Operations Research*, Vol. 42, No. 5, September–October 1994, pp. 793–813.

Keeney, Ralph L., "Structuring Objectives for Problems of Public Interest," *Operations Research*, Vol. 36, No. 3, May–June 1988, pp. 396–405.

Keeney, Ralph L., *Value-Focused Thinking,* Boston, MA: Harvard University Press, 1992.

Keeney, Ralph L., Detlof von Winterfeldt, and Thomas Eppel, "Eliciting Public Values for Complex Policy Decisions," *Management Science,* Vol. 36, No. 9, September 1990, pp. 1011–1030.

Keeney, Ralph, and Craig Kirkwood, "Group Decision Making Using Cardinal Social Welfare Functions," *Management Science,* Vol. 22, No. 4, December 1975, pp. 430–437.

Keeney, Ralph L., and Timothy L. McDaniels, "Value-Focused Thinking About Strategic Decisions at BC Hydro," *Interfaces,* Vol. 22, No. 6, November–December 1992, pp. 94–109.

Kirkwood, Craig W., "An Overview of Methods for Applied Decision Analysis," *Interfaces,* Vol. 22, No. 6, November–December 1992, pp. 28–39.

Kirkwood, Craig W., *Multiobjective Decision Analysis,* Department of Decision and Information Analysis, Arizona State University, 1995. (Forthcoming as *Strategic Decision Making,* Belmont, CA: Duxbury Press.)

Kirkwood, C. W., and R. K. Sarin, "Preference Conditions for Multiattribute Value Functions," *Operations Research,* Vol. 28, pp. 225–232, 1980.

Köksalan, M. Murat, and Paul N.S. Sagala, "Interactive Approaches for Discrete Alternative Multiple Criteria Decision Making with Monotone Utility Functions," *Management Science,* Vol. 41, No. 7, July 1995, pp. 1158–1171.

Mead, Lawrence M., Review of *Values and Public Policy,* in Henry J. Aaron, Thomas E. Mann, and Timothy Taylor (eds.), *Journal of Policy Analysis and Management,* Vol. 14, No. 2, pp. 327–359.

Pérez, Joaquin, "Some Comments on Saaty's AHP," *Management Science,* Vol. 41, No. 6, June 1995, pp. 1091–1095.

Quade, E.S., *Analysis for Public Decisions,* New York: American Elsevier, 1975.

Rostker, Bernard, and Harry J. Thie, *The Defense Officer Personnel Management Act of 1980, A Retrospective Assessment*, R-4246-FMP, Santa Monica, CA: RAND, 1993.

Saaty, Thomas L., "An Exposition of the AHP in Reply to the Paper 'Remarks on the Analytic Hierarchy Process'," *Management Science*, Vol. 36, No. 3, March 1990, pp. 259–273.

Saaty, Thomas L., "How To Make a Decision: The Analytic Hierarchy Process," *Interfaces*, Vol. 24, No. 6, November–December 1994, pp. 19–43.

Saaty, Thomas L., "Thoughts on Decision Making," *OR/MS Today*, April 1996, pp. 8–9.

Sackman, H., *Delphi Assessment: Expert Opinion, Forecasting, and Group Process*, R-1283-PR, Santa Monica, CA: RAND, 1974.

Shapiro, Zur, "Making Trade-offs Between Job Attributes," *Organizational Behavior and Human Performance*, Vol. 28, 1981, pp. 331–355.

Stillwell, William G., Detlof von Winterfeldt, and Richard S. John, "Comparing Hierarchical and Nonhierarchical Weighting Methods for Eliciting Multiattribute Value Models," *Management Science*, Vol. 33, No. 4, April 1987, pp. 442–450.

Stonebraker, Jeffrey S., et al., *Decision Analysis Consulting Projects*, Department of Mathematical Sciences, USAFA-TR-95-1, Colorado Springs, CO: USAF Academy, 30 June 1995.

Thie, Harry J., and Roger A. Brown, *Future Career Management Systems for U.S. Military Officers*, MR-470-OSD, Santa Monica, CA: RAND, 1994.

U.S. Department of Defense, *Report of the Department of Defense Ad Hoc Committee to Study and Revise the Officer Personnel Act of 1947* ("Bolte Report"), December 1960.

Weber, Martin, Franz Eisenfuhr, and Detlof von Winterfeldt, "The Effects of Splitting Attributes on Weights in Multiattribute Utility Measurement," *Management Science*, Vol. 34, No. 4, April 1988, pp. 431–445.

Winkler, Robert L., "Decision Modeling and Rational Choice: AHP and Utility Theory," *Management Science*, Vol. 36, No. 3, March 1990, pp. 247–273.